Motor Vehicle Work

Editor
R. Brooks
Senior Lecturer in Motor Vehicle Subjects
Bolton College of Education (Technical)

Author
J. Whipp
Lecturer in Motor Vehicle Subjects
Moston College of Further Education

M

First edition 1973
Reprinted 1974, 1975, 1979

Published by
THE MACMILLAN PRESS LTD
London and Basingstoke
Associated companies in Delhi Dublin
Hong Kong Johannesburg Lagos Melbourne
New York Singapore and Tokyo

ISBN 0 333 14424 4

Printed in Great Britain by
Lowe & Brydone Printers Ltd, Thetford, Norfolk

PREFACE

This book is specifically intended for young people in about their last year at secondary school. Its aim is to enable them to capitalise on their natural interest in becoming individually mobile at a time when frustration and boredom can—if not prevented—all too easily creep in.

ACKNOWLEDGEMENTS

The editor and author wish particularly to thank the following firms for their invaluable help with the illustrative matter in this book.

AC-Delco, Division of General Motors Ltd
British Leyland Motor Corporation Ltd
Castrol Ltd
Dunlop Ltd
Fiat (England) Ltd
Girling Ltd
Honda Ltd
Joseph Lucas Ltd

Renault Ltd
Vauxhall Motors Ltd

CONTENTS

INTRODUCTION

Careful thought has been given to the progressive structure of the material herein. It is presented in convenient units, in a manner neither too infantile to be boring nor too advanced not to be easily understood and assimilated. At all times the likely limitations of equipment has been borne in mind. Safety aspects are dealt with naturally, where they arise, in context.

The exercises, vehicle investigations and project work are all designed to be interesting and rewarding, and most can be tackled to varying depths according to individual preference. For those hoping to take up a career in automobile engineering, this could provide a good 'taster' and basis, without encroaching on their later trade training; for all, however, it should provide them with a greater understanding of one of the dominant features of modern life: the motor vehicle.

SECTION 1
VEHICLE LAYOUT

The layout of a vehicle concerns the arrangement of the main components. The following items are generally accepted as being the main components:

Engine, clutch, gearbox, final drive, suspension

One of the most popular arrangements is known as the 'orthodox' or 'conventional' arrangement, this is shown in simplified form below.

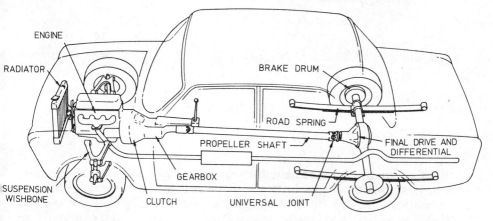

In the arrangement shown above, the engine is at the front of the vehicle and the drive is to the rear wheels. Three advantages offered by this arrangement are:

(a) Simplicity.
(b) Load distributed fairly evenly about the vehicle frame.
(c) Ease of maintenance (components can be removed individually for repair or replacement).

Two alternatives to this arrangement are:

Front-mounted engine driving front wheels.

Rear-mounted engine driving rear wheels.

The arrangement shown below is that of a rear engine rear wheel drive vehicle. One feature of this arrangement is that the final drive is positioned in between the engine and gearbox, although the drive is taken from the engine to the clutch and gearbox and then into the final drive.

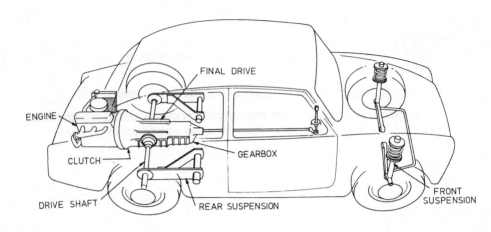

An alternative arrangement for rear engine rear wheel drive vehicles, is to position the engine in front of the rear axle or final drive, that is, immediately behind the seating compartment. This is shown below and is known as the 'mid-engine' arrangement, it is very often employed on sports and racing cars.

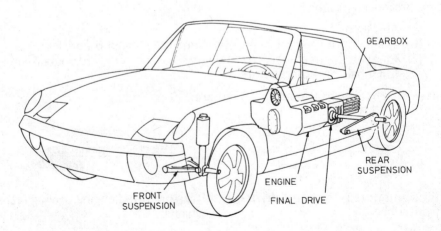

EXERCISE 1. LAYOUT

Questions

(1) List and describe the function of five important vehicle components, other than those already mentioned.

(2) Why do you think the arrangement shown on page 1 is referred to as the conventional or orthodox arrangement?
Name four makes and models of modern vehicles using this arrangement.

(3) Name two makes and models of modern vehicles using the front engine, front-wheel-drive layout, and two makes and models of modern vehicles using the rear-engine, rear-wheel-drive layout.

(4) One advantage of the two alternative layouts mentioned in question (3) is that the weight is concentrated on the driving wheels.
Give one other advantage and one disadvantage for each layout.

(5) Why is it convenient for sports and racing cars to adopt a mid-engine layout?

Vehicle Investigation (State make, model and year of vehicle)

(a) Sketch the layout of the main components of a sports or racing car.

(b) Examine a vehicle of non-conventional layout and comment on the accessibility of the major parts from the repair and maintenance viewpoint.

(c) Comment on how the positions of the main components affect the overall body shape and styling.

(d) Sit in the driver's seat with the safety belt correctly (i.e. fairly firmly) fastened. Check whether or not you are able to comfortably reach all the controls and accessories such as radio and glove box. List those which cannot be properly operated.

Project Work

(i) Front-wheel-drive cars are becoming very popular. List as many makes and models as you can that use this arrangement and state what are its advantages.

(ii) Look at some of the most modern vehicles in current production and illustrate what you consider are their most important features. (Note: vehicle sales brochures are especially helpful for this type of project.)

SECTION 2
THE PETROL ENGINE

FUNCTION

The engine provides the power to turn the vehicle's wheels and to drive auxiliaries such as the dynamo, water pump and perhaps power-steering pump and the like.

TYPES OF ENGINE

The engines used in motor vehicles are (almost always) heat engines and, as the name implies, rely upon heat to generate power. The petrol engine, which is employed in most cars, is a form of heat engine. It is called the 'internal combustion' engine. There are a few vehicles driven by other means such as electricity and steam, but they are extremely rare.

PRINCIPLE OF OPERATION

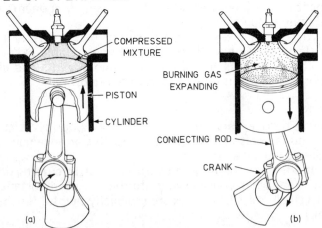

The drawing at (a) above shows a mixture of petrol and air which has been compressed into a confined space by a piston moving upwards in an enclosed cylinder. The mixture is then ignited and the heat generated due to burning causes the gas to expand rapidly; so forcing the piston down the cylinder as shown at (b). The connecting rod, attached to the piston at the top and the crank at its lower end, converts the downward movement of the piston into a rotary movement. As shown at (b) the reciprocating (up and down) motion of the piston is converted into the rotary motion of the crankshaft by the action of the connecting rod.

THE FOUR STROKE CYCLE

The majority of motor vehicle engines operate on what is known as the 'four stroke' or 'Otto' cycle; a stroke being the movement of the piston from top to bottom of the cylinder, or vice versa. The four strokes, in order of sequence, are: INDUCTION, COMPRESSION, POWER and EXHAUST, these are described below.

Induction stroke

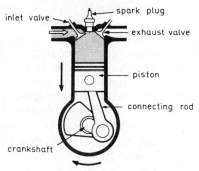

Compression stroke

The piston moves down the cylinder creating a partial vacuum which draws in a mixture of petrol and air through the open inlet valve.

Both valves are closed and the piston moves up the cylinder to compress the gas.

Power stroke

Exhaust stroke

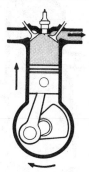

An electric spark ignites the mixture and the piston is forced down the cylinder; both valves are still closed.

The exhaust valve is opened and the piston moves up the cylinder so pushing the burnt gases out through the exhaust system. The cycle is then repeated.

5

THE TWO-STROKE CYCLE

The two-stroke cycle enables the four operations, i.e. induction, compression, power and exhaust to be completed on only two strokes of the piston. This is achieved by the use of 'ports' in the cylinder wall which are covered and uncovered by the piston as it moves in the cylinder; that is, the piston itself acts as a type of valve.

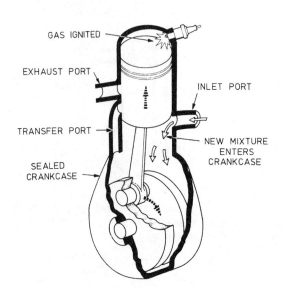

GAS IGNITED

EXHAUST PORT

INLET PORT

TRANSFER PORT

NEW MIXTURE ENTERS CRANKCASE

SEALED CRANKCASE

Operation

It can be seen from the drawing opposite, that events occur both above and below the piston. As the piston approaches the top of the cylinder, the mixture above it is compressed. At the same time below the piston new mixture is drawn into the crankcase through the inlet port.

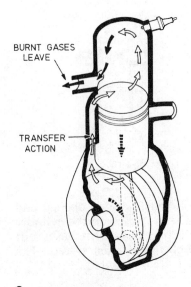

BURNT GASES LEAVE

TRANSFER ACTION

When the mixture in the combustion chamber is ignited by the spark, the piston is forced down the cylinder. Towards the end of the stroke, the piston uncovers the exhaust port and some of the exhaust gases leave the cylinder. As can be seen from the drawing opposite, the transfer port is also uncovered and new mixture which has been lightly compressed by the descending piston enters the cylinder. At this stage the remaining exhaust gas is dispelled by the new mixture as it enters the cylinder.

MULTI-CYLINDER ENGINES

One disadvantage of a *single* cylinder, four-stroke engine is that the power or working stroke occurs only once every *two* engine revolutions. The turning effort (torque) produced is therefore uneven and must be smoothed out with the aid of a large, heavy 'flywheel' attached to the output shaft (crankshaft). This problem of uneven torque output, as well as other problems associated with large single cylinder engines, does therefore make them unsuitable for use in the majority of cars.

By increasing the number of cylinders, the engine will run more smoothly and produce more power. It can be seen from the drawings below, that the firing intervals of a *four*-cylinder engine can be arranged to give a power stroke every 180° (that is, half revolution) of crankshaft rotation. This then gives two power strokes per engine revolution.

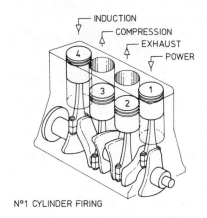

N°1 CYLINDER FIRING

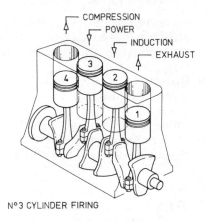

N°3 CYLINDER FIRING

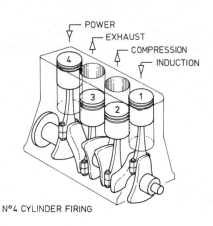

N°4 CYLINDER FIRING

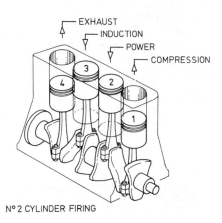

N°2 CYLINDER FIRING

EXERCISE 2A. ENGINE OPERATION

Questions

(1) Why is the piston forced down the cylinder when the petrol/air mixture is burned?

(2) How many engine revolutions take place between the succeeding power strokes of a single cylinder four-stroke engine?

(3) What causes the mixture to: (a) enter the crankcase of a two-stroke engine; (b) transfer into the cylinder?

(4) State two advantages of a two-stroke engine when compared with a four-stroke engine.

(5) What is the firing order of the four-cylinder layout shown on the previous page?

Vehicle Investigation

Complete the following by examining a multi-cylinder engine. (State make, model and year of vehicle.)

(a) State the engine type, capacity and number of cylinders.

(b) Remove the valve cover and identify the valves (as inlet or exhaust) by observing their positions relative to the inlet and exhaust manifold.

(c) Rotate the engine* and determine the firing order by observing the valve movement.

(d) Examine a small motor cycle or scooter and state the engine type, capacity and number of cylinders.

(e) Sketch the engine and show the position of the carburettor and exhaust outlet on the cylinder assembly.

Project Work

(i) Determine who made the first practical four-stroke internal combustion engine and trace its development. (The name 'Otto' is a useful starting point.)

(ii) In the early years some vehicles were driven by steam and some by electricity. Recently motor car manufacturers have revived their interests in these types of power units. What are the main reasons for their renewed interest? (General Motors and British Leyland have done much work in this field.)

---

* Where no starting handle is available it may well be necessary to remove the sparking plugs and turn the engine by pulling on the fan blades. The ignition must be *switched off*. Do not rotate the engine by the starter.

MAIN ENGINE COMPONENTS

The components which make up a complete engine are shown in the drawing below and the functions of some of the main components are stated in the table.

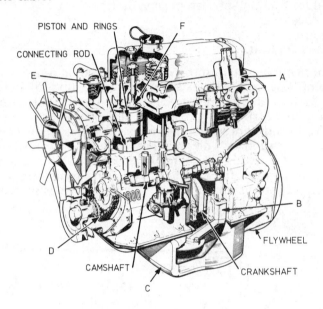

COMPONENT	FUNCTION
Flywheel	To damp out engine speed and torque fluctuations due to the effect of the firing impulses. To help the engine over its 'idle' strokes. It also provides a mounting for the clutch and starter ring gear.
Piston and rings	To form a gas-tight seal in the cylinder and to transmit the force of the expanding gas to the connecting rod.
Connecting rod and crankshaft	To convert the 'linear' (reciprocating) movement of the piston into a rotary movement at the crankshaft.
Camshaft	This shaft rotates at half the speed of the crankshaft and it 'cams' open the valves at the appropriate time in the cycle.

ENGINE LUBRICATION

The lubricating oil in an engine serves a number of purposes; it helps greatly to:

(a) reduce friction between moving parts;
(b) reduce wear;
(c) reduce noise;
(d) dissipate heat.

The lubrication system most widely adopted on cars is known as the 'wet sump' system. In this system the lubricating oil is carried in a reservoir (sump) at the base of the engine. An oil pump (usually driven from the camshaft) picks up the oil and delivers it, under pressure, through various drillings and pipes to certain parts of the engine. Whilst the engine is running, oil splashes around vigorously inside the crankcase, and it is this splashing which serves to lubricate most other parts of the engine: for example, cylinders, cams and followers. A typical lubrication system for a four cylinder engine is shown below.

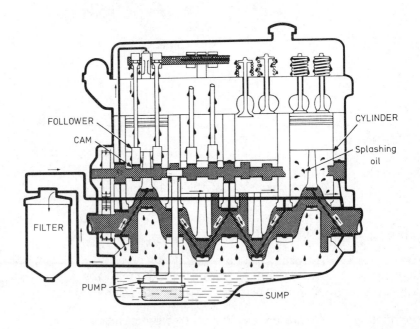

The Oil Pump

A simple and popular type of oil pump is shown at the top of page 11. This type of pump is usually submerged in the oil and as the gears rotate the oil is trapped between the casing and the tooth cavities. When the gears mesh, the oil, as shown in the sketch, is forced out on the delivery side.

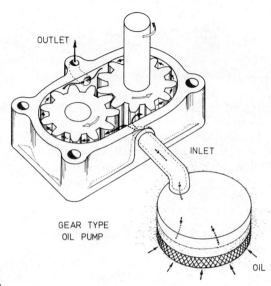

OUTLET

INLET

GEAR TYPE
OIL PUMP

OIL

Oil Filtration

It is usual for the oil on its way from the sump to the bearings to pass through two filters. The first (or primary) filter is made of relatively coarse wire mesh and prevents any large particles entering the pump. The second (secondary) filter is normally in the pipe line between the pump and the bearings and this is a very fine filter.

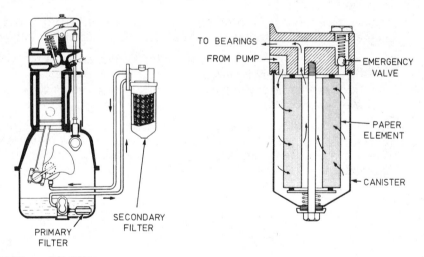

TO BEARINGS
FROM PUMP
EMERGENCY VALVE

PAPER ELEMENT

CANISTER

PRIMARY FILTER

SECONDARY FILTER

Full-Flow Oil Filter

The principle of operation of this typical secondary filter is shown above. Oil enters the canister from the pump and the build-up of pressure in the canister forces the oil through the paper element. In the event of the filter becoming clogged, the increase in oil pressure will open the emergency valve and the oil will by-pass the filter.

11

EXERCISE 2B. LUBRICATION

Questions

(1) Make a well-proportioned sketch of a connecting rod and label the main parts.
(2) Name the lettered parts on the engine shown on page 9.
(3) How is a two-stroke petrol engine usually lubricated?
(4) Study the lubrication system shown on page 10 and name the engine parts which are 'pressure fed' with lubricant.
(5) Name six engine parts lubricated by oil-splash.

Vehicle Investigation

(State make, model and year of vehicle)

(a) Remove the valve cover and make a simple sketch to show how the valves are made to open.
(b) Sketch a simple rectangle to represent the cylinder-head and (as if looking down on to the head) show the relative positions and respective shapes of the inlet and exhaust manifolds.
(c) Make simple sketches to show where the secondary (external) filter, the oil filler and the dip-stick are situated on the engine.
(d) How is the driver given an indication of oil pressure?
(e) What are the recommended engine oil-change intervals?

Project Work

(i) Choose four current motor cars, each having different types of engines and give the following information about each:

     (a) engine type;   (e) engine capacity;
     (b) number of cylinders;   (f) power output;
     (c) cylinder arrangement;   (g) price of vehicle.
     (d) firing order;

(ii) Motor car engines are made up using different metals in different places. List some of the main components, give examples of the type of metal used and explain why it is used in that particular application.

SECTION 3
THE IGNITION SYSTEM

PURPOSE

The purpose of the ignition system is to create an electric spark in the engine combustion chamber, at exactly the right time, which will ignite the mixture of petrol and air.

MAIN COMPONENTS

The ignition system is made up of the following main components:

Battery, ignition switch, coil, contact points, condenser, distributor, sparking plugs and cables.

The main components which are obviously associated with the ignition system are shown below.

The coil is normally located in the engine compartment on a body or engine bracket close to the distributor. Its purpose is to transform the low voltage from the battery to a high voltage for producing a spark.

The distributor is located on the engine and has a central shaft which is driven at half engine speed from the camshaft. Its purpose is to act as an automatic switch in the low tension circuit and to direct the electricity from the coil to the correct sparking plug at exactly the right time.

The sparking plug or plugs, in the case of a multi-cylinder engine, are screwed into the cylinder head and protrude slightly into the combustion chamber. The high tension current must jump the gap at the points and in so doing creates a spark which ignites the mixture.

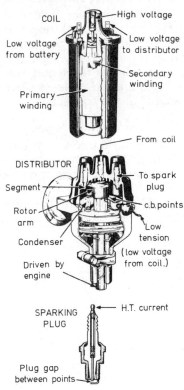

COIL — High voltage
Low voltage from battery
Low voltage to distributor
Secondary winding
Primary winding

From coil
DISTRIBUTOR
To spark plug
Segment
c.b.points
Rotor arm
Low tension (low voltage from coil.)
Condenser
Driven by engine

SPARKING PLUG — H.T. current
Plug gap between points

IGNITION CIRCUIT

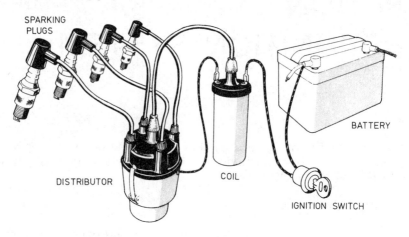

SPARKING PLUGS

BATTERY

DISTRIBUTOR

COIL

IGNITION SWITCH

OPERATION

Low Tension (Points closed)

With the ignition switched on and the contact points closed, current (at battery voltage) flows through the primary windings and through the contact points to earth. This current flow creates a magnetic field around the primary winding. As the engine rotates the distributor cam opens the contacts and the current ceases to flow. (The condenser ensures a rapid collapse of the magnetic field and in doing so prevents 'arcing' at the contacts.)

High Tension (Points open)

Stopping the low tension current flow causes a rapid collapse of the magnetic field. This collapse induces a high voltage into the secondary winding which forces a current along the high tension lead to the rotor arm, which at that moment is in line with a segment in the distributor cap. The current is then directed from the segment, along the lead to the sparking plug; where it jumps the gap and creates the spark.

EXERCISE 3. IGNITION

Questions

(1) Why are the leads from the distributor cap much thicker than the other cables in the system?
(2) What component changes the low voltage battery current into the high voltage current needed at the spark plugs?
(3) Why is a high voltage current needed at the spark plugs?
(4) What component controls the timing of the ignition spark?
(5) What is the function of the contact breaker points and why is the 'heel' of the moving point made of fibre?

Vehicle Investigation (State make, model and year of vehicle)

(a) How many cylinders has the engine and what is their firing order?
(b) Carefully measure the gaps of the sparking plugs and contact breaker points. Tabulate the results and check them against those recommended in the vehicle handbook.
(c) Examine the ignition wiring and terminals for damage, deterioration or looseness and list any defects discovered.
(d) Remove, clean (or replace) and reset the gap of the contact breaker points and/or the spark plugs.
(e) Determine how current is transferred from the centre lead of the distributor to the rotor arm.

Project Work

(i) Transistorised ignition is now becoming very popular. Find out how it works and its advantages.
(ii) Diesel (or compression ignition) engines are very similar to petrol engines, but need no ignition system. Find out how they work.

SECTION 4
THE FUEL SYSTEM

SAFETY—REMEMBER PETROL VAPOUR IS HIGHLY FLAMMABLE

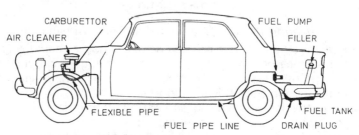

The fuel tank is nearly always at the opposite end of the car to the engine for reasons of safety, space availability and weight distribution. Tanks are usually made of mild steel and their fuel capacity is normally sufficient to allow the car to travel about 300 miles without refuelling. It is important that every tank must be provided with a vent to atmosphere.

PETROL PUMP

Most cars employ a mechanical type fuel pump, mounted on the side of the engine and driven by the camshaft, in order to lift the fuel from the low-mounted tank to the high-mounted carburettor. In some instances however, an electrically operated pump is used and this may be fitted in almost any position between the tank and the engine.

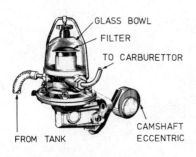

AC MECHANICAL PUMP

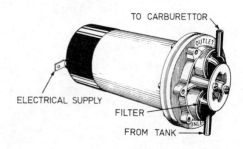

SU ELECTRIC PUMP

16

AIR FILTERS (Cleaners)

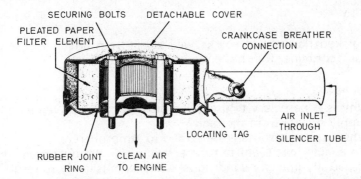

SECURING BOLTS   DETACHABLE COVER

PLEATED PAPER
FILTER ELEMENT

CRANKCASE BREATHER
CONNECTION

AIR INLET
THROUGH
SILENCER TUBE

LOCATING TAG

RUBBER JOINT   CLEAN AIR
RING   TO ENGINE

Most modern vehicles are fitted with replaceable paper-element type air cleaners. The air going into the engine via the carburettor must pass through the specially impregnated element which retains all grit and dust particles of harmful size. Such filter elements require to be replaced at intervals of about 10 000 to 12 000 miles, otherwise insufficient air may be able to enter the engine and fuel consumption will increase.

Another important job of the filter is to silence the air going into the engine. Without an air filter, the air entering the carburettor creates an annoyingly loud sucking noise.

A simple refinement on some air cleaners is that the air-inlet tube has a summer and winter position. For example, drawing air from near a hot part of the engine in winter and from a relatively cool part of the under-bonnet in summer.

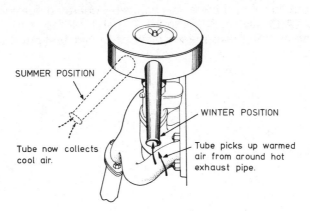

SUMMER POSITION

WINTER POSITION

Tube now collects
cool air.

Tube picks up warmed
air from around hot
exhaust pipe.

17

EXERCISE 4A. FUEL SYSTEM and AIR FILTERS

SAFETY

Keep naked flames and sparks well away from any petrol and away from any container that has previously held petrol.

Questions

(1) Why is it necessary to use a flexible pipe near the engine in the fuel supply system from the tank?
(2) Why must a petrol tank be vented to atmosphere?
(3) An engine may seem to run perfectly well without an air filter, but (especially in dusty conditions) what harm is likely to result?
(4) How should you deal with a minor petrol fire?

Vehicle Investigation (State make, model and year of vehicle)

(a) State what type of fuel pump is fitted and make a sketch to show its position on the car.
(b) By means of a sketch show how the fuel tank is vented to atmosphere.
(c) Sketch the air cleaner and show how it fits to the engine. Examine the filter element and comment on its condition.
(d) Run an engine with and without an air cleaner. Assess the difference in noise level.

Project Work

(i) Determine how a motorist can accurately check the fuel consumption of his vehicle.
(ii) Find out in detail how a mechanically operated fuel pump works. (Note: A.C. Delco Division of General Motors Ltd., Stag Lane, London N.W.9, make the majority of British fuel pumps.)

BASIC CARBURATION

Curiously enough, although one tends to think of petrol burning inside the engine in order to produce power, it is a fact that petrol, on its own, will not burn; it must be supplied with oxygen (that is, in practice mixed with air, which contains oxygen). It is the function of the carburettor to mix the air and petrol in correct proportions according to the engine requirements; and by means of the throttle valve (linked to the accelerator pedal) enable the driver to vary the quantity of mixture supplied to the engine, thus regulating the power output.

In spite of the fact that carburettors are precision-made instruments and look very complicated, they very rarely go wrong. Their basic operating principle is shown by the diagram below.

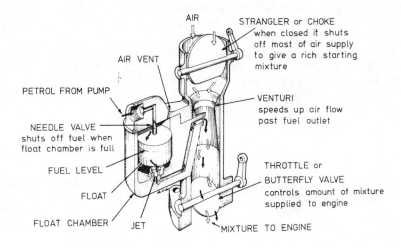

AIR

STRANGLER or CHOKE
when closed it shuts
off most of air supply
to give a rich starting
mixture

AIR VENT

PETROL FROM PUMP

VENTURI
speeds up air flow
past fuel outlet

NEEDLE VALVE
shuts off fuel when
float chamber is full

FUEL LEVEL

THROTTLE or
BUTTERFLY VALVE
controls amount of mixture
supplied to engine

FLOAT

FLOAT CHAMBER   JET

MIXTURE TO ENGINE

AIR/FUEL MIXTURE

Normally motorists buy petrol in gallon units—measures of volume. However, when considering carburation the proportion of air to petrol is usually given by weight. The chemically correct mixture strength needed to obtain complete combustion (burning) is 15 pounds of air to 1 pound of petrol. The following are examples of typical mixture strengths for various conditions encountered in service.

Starting-engine cold	9 : 1
Idling or 'tick-over'	12 : 1
Acceleration	12 : 1
Economy	16 : 1
Maximum power	12 : 1

The simple carburettor shown on the previous page could work satisfactorily on a single speed engine (such as say a concrete mixer) but could not cope with the sort of demands listed above. This is because the depression (or 'suction') in the fixed-size venturi varies enormously according to the speed of the air passing through it, since the depression here governs the amount of petrol drawn from the carburettor. For this reason some carburettors have several jets and flow systems.

Constant Depression Carburettor

(Also known as constant vacuum- variable jet carburettor)

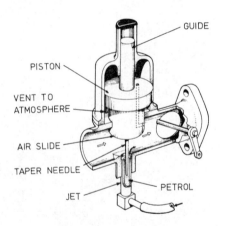

This type of carburettor is so made that the depression over the jet is kept constant by means of lifting the air slide and piston; whilst the *effective* size of the jet is varied by the taper needle.

EXERCISE 4B. CARBURATION

SAFETY

Exercise great caution to ensure that petrol vapour cannot become ignited.

Questions
(1) When operating the choke during cold starting, what effect does it have on carburation?
(2) What would be the effects of running an engine for prolonged periods 'on choke'?
(3) What keeps the fuel in the float chamber at a constant level?
(4) In the 'constant depression' type carburettor how is the *effective* size of the jet altered? Show this by means of a sketch.

Vehicle Investigation (State make, model and year of vehicle)
(a) State the make and type of carburettor fitted and make a sketch to show the location of the points at which adjustment may be made.
(b) Examine the carburettor and fuel system for any signs of leakage or other faults. List all the parts inspected and state their condition.
(c) Ideally the nose of the centre electrode of the spark plugs should be a light, chestnut brown and the inside of the exhaust tail pipe a light grey. This denotes a correct mixture strength. Report on the condition of these parts on the car being inspected. (Note: These colours do not apply to an engine that has been subjected to longish periods of idling.)

Project Work
(i) Study one make of carburettor and determine in detail how it works. (Note: The Zenith Carburettor Co. Ltd., Stanmore, Middlesex, are Britain's largest manufacturers of carburettors.)
(ii) Find out the chemical constituents of petrol.

SECTION 5
THE COOLING SYSTEM

THE COOLING SYSTEM

During combustion the temperature of the burning gases are (momentarily) sufficient to melt the metal of the cylinder. It is the function of the cooling system to prevent the engine becoming overheated (in practice the oil film would be burned away and the pistons expand and seize in position before the engine melted), but at the same time to allow it to operate at a temperature high enough to ensure efficient operation.

WATER COOLING

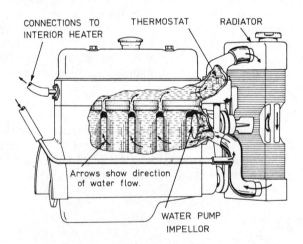

CONNECTIONS TO INTERIOR HEATER   THERMOSTAT   RADIATOR

Arrows show direction of water flow.

WATER PUMP
IMPELLOR

Although water can be made to circulate naturally by convection currents, cooling systems fitted with pumps (or impellors) to assist and direct water flow are now virtually universal. The water conducts the heat away from the hot parts of the engine and is itself cooled, as it passes through the radiator, by giving up its heat to the surrounding air.

THERMOSTAT

A thermostat is a temperature-sensitive valve usually fitted in its own housing on the cylinder head *(see diagram on previous page) and* controls the water flow from the engine to the radiator. Its function is to enable the engine to warm-up quickly by restricting the flow in the cooling system; and then to control water flow so that the engine is running at its optimum temperature. This is likely to result in a coolant temperature of between 75° C and 85° C.

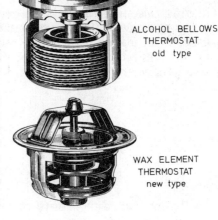

ALCOHOL BELLOWS
THERMOSTAT
old type

WAX ELEMENT
THERMOSTAT
new type

RADIATOR PRESSURE CAP

The radiator cap plays a most important function in keeping the pressure within the cooling system several pounds per square inch above atmospheric pressure. Some of the resulting benefits are:

(a) Boiling point of coolant is raised.
(b) Evaporation is eliminated.
(c) Less topping-up required.
(d) No loss of coolant due to surge such as when cornering quickly.
(e) Radiator temperature is higher and so it dissipates more heat. Consequently size, weight and cost are reduced.
(f) Scale deposits are slow to form.
(g) Water-pump efficiency is maintained.

Note the warning:
'REMOVE CAP SLOWLY'
or in some cases,
'REMOVE ONLY WHEN COLD'

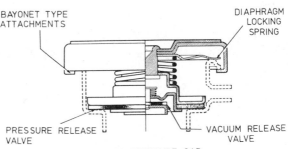

BAYONET TYPE
ATTACHMENTS

DIAPHRAGM
LOCKING
SPRING

PRESSURE RELEASE
VALVE

VACUUM RELEASE
VALVE

TYPICAL PRESSURE CAP
shown in half section

AIR COOLING

The fins, cast integral with the cylinder barrel, provide the necessary large
surface area needed to facilitate rapid
heat dissipation. Notice that the fins
are largest where the highest temperatures occur.

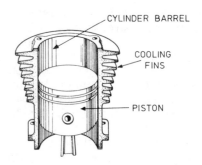

Although the vast majority of motor cycles are air-cooled, this form
of cooling is used on relatively few cars and those are almost always
small ones. Continental manufacturers who do use it include Citroen,
Fiat and, of course, Volkswagen.

The biggest advantage of air cooling is its simplicity and therefore
cheapness of manufacture; additionally the engine tends to warm-up
quickly and there is, obviously, no radiator, water or antifreeze
required. However, since there is no water jacket surrounding the
engine it tends to sound noisier and the interior heaters of air-cooled
cars are often not particularly satisfactory.

When fitted into a car an air-cooled
engine needs a large fan, which
absorbs quite a lot of power, to
direct a sufficient air flow around
the cooling fins.

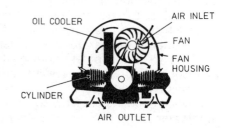

EXERCISE 5   COOLING

Questions

(1) Why do the very high temperatures that occur inside the engine cylinders not cause the cooling water to boil?

(2) Natural convection currents could be used to obtain water circulation without the expense and complication of a water pump; for what reasons is it not used on modern cars?

(3) What are the disadvantages of water cooling?

(4) What would be likely to happen if an engine thermostat were to fail in the closed position?

(5) Why is it recommended that radiators and cooling fins are not made shiny, but given a dull, black finish?

Vehicle Investigation (State make, model and year of vehicle)

(a) What is the normal operating temperature of the engine coolant?

(b) Using equipment similar to that shown, determine the opening and closing temperatures of the thermostat.

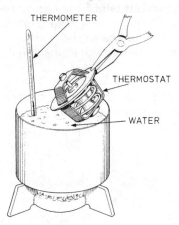

(c) Check how the results of (b) compare with the manufacturers' specification. (Note: Look carefully for temperature marking on thermostat.)

(d) Carefully inspect a cooling system for defects and list all items that require attention.

Project Work

(i) Why is it necessary to add antifreeze to water in frosty conditions and of what does antifreeze consist?

(ii) Find out details of the makes and models of cars currently available, that are fitted with air-cooled engines.

SECTION 6
THE TRANSMISSION

THE CLUTCH

The clutch is a coupling (normally of the friction type) through which the drive from the engine to the gearbox is transmitted.

Function
(a) To allow a gradual take-up of the drive from the engine, which may be rotating at say 1 000 rev/min, to the transmission and driving wheels which may be stationary.
(b) To provide a means of temporarily disconnecting the drive from the engine to the gearbox.
(c) To assist in gear changing.

Principle of Operation
The drawings below represent a much simplified clutch assembly in the engaged (driving) position and in the disengaged (non-driving) position.

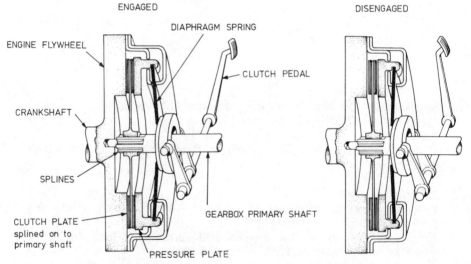

Operation
To move the vehicle off from rest, the driver will start the engine, depress the clutch pedal (to disengage the clutch) and engage low gear.

The clutch pedal is then gradually released until the spinning pressure plate and flywheel assembly begin to clamp the stationary clutch plate between them.

Slipping will take place between the friction linings and the pressure plate and flywheel assembly until full clamping pressure is achieved; that is, when the clutch pedal is fully released. It is this slipping action which gives the gradual take up of the drive.

Construction

The main components that make up the clutch and its operating mechanism are shown below, the clutch itself being shown 'exploded' to reveal the clutch or centre plate.

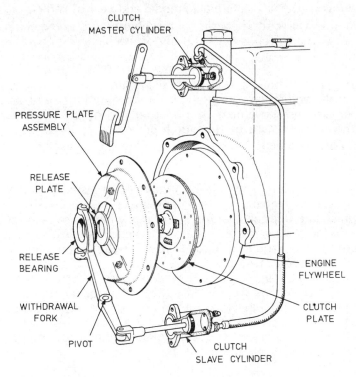

The clutch assembly shown above is only one of many types in use. The clamping pressure of the type shown is provided by a number of coil springs in the pressure plate assembly. One other type of clutch in widespread use is the diaphragm spring clutch which is similar to that shown above, but uses only one large saucer-shaped spring. Some clutches, especially those on motor cycles are of the 'multiplate' type; that is, they have a number of friction faces.

Automatic clutches vary considerably and may use, for example, fluid (oil) or magnetised powder to take up the drive.

EXERCISE 6A   THE CLUTCH

Questions

(1) Why is it necessary that a driver must be easily able to disconnect the drive from the engine to the gearbox? (Consider only a conventional type of vehicle.)
(2) Why is special friction material required on the clutch plate?
(3) What is meant by the term 'clutch slip'?
(4) Some means of adjustment, either manual or automatic, must be provided in the clutch-operating mechanism. Why is this necessary?
(5) Considering the operation of a clutch; what type of driving habits could seriously reduce the life of the clutch?

Vehicle Investigation (State make, model and year of vehicle)

(a) Examine the clutch-operating mechanism on a vehicle and state whether the clutch is hydraulically or mechanically operated.
(b) Show by a sketch how clutch adjustment is carried out on the vehicle.
(c) Does the mechanism require any immediate maintenance; for example, adjustment, 'topping up' with hydraulic fluid, and so on?
(d) Compare the movement of the outer end of the withdrawal lever with the movement of the clutch pedal and state the approximate ratio of movement.

Project Work

(i) The clutch plate itself looks to be of very simple construction. However, in fact it is quite a complex component with some very sophisticated design features. Find out what these features are and what they do. (Note: The Borg and Beck Co. Ltd., of the Automotive Products Group, Leamington Spa, are probably Britain's largest makers of the plates.)
(ii) Before World War II very few cars were fitted with automatic clutches (Daimlers were in fact the first to be so fitted). However today such clutches are quite commonplace and indeed must be fitted to cars with automatic transmissions. List some of the modern cars using automatic clutches; find out how they work and their advantages and disadvantages.

THE GEARBOX

A purpose of the gearbox is to allow the driver (by selecting the appropriate gear) to vary the turning effort (torque) applied to the driving wheels according to the 'speed' and 'load' requirements.

The engine can only produce a certain amount of torque (turning effort). By engaging low gear this torque is increased, thus delivering more torque to the driving wheels. However, the speed of the wheels is decreased in the same proportion as the torque is increased.

When the vehicle is cruising easily along a level road, the load is relatively light; thus the driver can select high gear which may well give no torque multiplication and obviously no reduction in speed between the engine and transmission. Consequently road speed will be much higher for a given engine rev/min than it would be in low gear.

The gearbox also provides a means of reverse; and a permanent position of neutral.

Principle of Operation

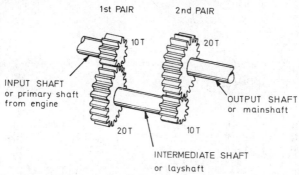

Consider the gear train shown above. The turning effort applied by the input shaft will be doubled by the first pair of gears, owing to the fact that the larger gearwheel has twice the number of teeth (and thus twice the radius) of the smaller gearwheel. This effort, which is now applied to the second similar pair of gears by the layshaft, is again doubled. The turning effort at the output shaft will therefore be four times (that is 2 × 2) that at the input shaft, although the speed of the output shaft will be four times less (that is, one quarter) that of the input shaft.

Simple Motor Car Gearbox (3-speed sliding mesh type)

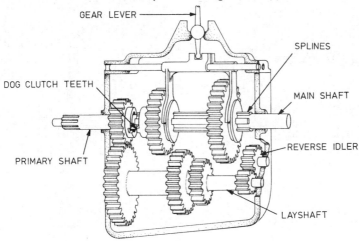

29

In the type of motor car gearbox shown on page 29, the different speeds are obtained by connecting, in turn, the mainshaft gearwheels to different sized gearwheels on a layshaft, which is in 'constant mesh' with the input or 'primary' shaft. Top gear is a direct drive (1 to 1) through the gearbox and is obtained by joining the input and output shafts by means of a 'dog-clutch' connection.

The gear positions and 'power paths' in each gear are shown below for a three-speed and reverse gearbox.

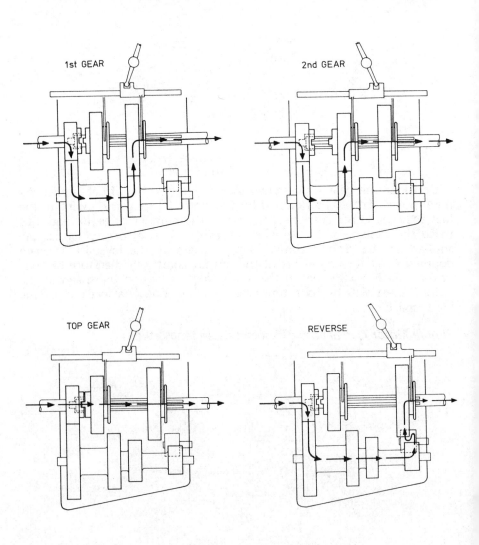

Constant-Mesh Type Gearbox

Whilst the sliding-mesh type of gearbox just mentioned is relatively simple and cheap, it has certain disadvantages, e.g.:

(a) Gearchanging is difficult, i.e. the driver must synchronise the speeds of the meshing gearwheels as the gearchange takes place, this calls for skill and experience.

(b) The type of gear teeth (i.e. straight-cut 'spur' teeth) are noisy.

To overcome these disadvantages most modern vehicles employ the constant-mesh type of gearbox.

Principle of Operation. (Simplified layout)

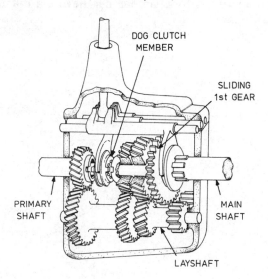

As can be seen from the drawing above, all the gearwheels are in 'constant mesh', but with the mainshaft wheels being free to spin on the mainshaft. When a gear is engaged, the 'dog-clutch' members, which are themselves splined to the mainshaft, connect the appropriate gearwheel to the mainshaft; drive is then transmitted through the 'box in exactly the same way as in the sliding-mesh type mentioned on the previous page.

Synchromesh Gearbox

In modern gearboxes, the sliding dog-clutch members are part of a SYNCHROMESH DEVICE, which, as the name implies automatically synchronises the speeds of the meshing dog-clutches during gear-changing, thereby making gear changing much easier. For obvious reasons wear and tear on the gears is less, and due to the design of the gearteeth the gearbox is quieter in operation than the sliding-mesh type.

31

AUTOMATIC TRANSMISSION

In cars fitted with a conventional, manually controlled clutch and gearbox, their operation requires dexterity and effort on the part of the driver. When a car is fitted with automatic transmission the driver no longer needs to operate the clutch or change gear. This not only contributes towards easier (and thus safer) driving, but also can substantially reduce wear and tear on the engine and the transmission itself.

The main drawbacks of this form of transmission are the relatively high first cost and the amount of power it absorbs from the engine. This last mentioned usually results in a loss of about two miles per gallon when compared with a similar car having a manually operated clutch and gearbox.

Simplified arrangement:

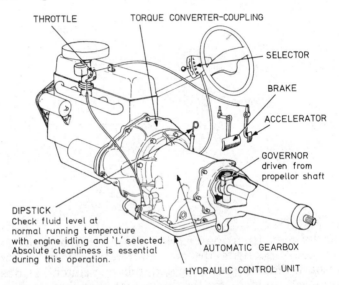

THROTTLE   TORQUE CONVERTER–COUPLING

SELECTOR

BRAKE

ACCELERATOR

GOVERNOR
driven from
propellor shaft

DIPSTICK
Check fluid level at
normal running temperature
with engine idling and 'L' selected.
Absolute cleanliness is essential
during this operation.

AUTOMATIC GEARBOX

HYDRAULIC CONTROL UNIT

After forward gear selection (D), as the engine speed is increased the fluid clutch automatically takes up the drive and the vehicle accelerates in low gear. Gear changing, up or down, takes place automatically due to the action of the hydraulic control unit. The engine speed and road speed at which the gear changes are made depends upon the accelerator position and the governor speed.

EXERCISE 6B. THE GEARBOX

Questions

(1) Considering the layout of both the sliding-mesh and constant-mesh gearbox, which shafts will always be turning when the engine is running and the gearbox is in neutral?

(2) Why is there little wear and tear on the gears themselves with a constant-mesh gearbox?

(3) Reverse is obtained by transmitting the drive from the layshaft to the mainshaft via an idler gear. Make simple sketches to show how this motion is achieved.

(4) Which gears in a sliding-mesh gearbox are always in mesh whatever gear or neutral position is engaged?

(5) Why is first gear in a constant-mesh gearbox sometimes a sliding-mesh gear?

Vehicle Investigation (State vehicle make and model)

(a) How many forward speeds has the gearbox?

(b) Study the gear lever movement for all gear positions and make a simple line diagram to show the gear lever position in each gear.

(c) By counting the number of turns of the engine to one turn of the propellor shaft (in each gear), determine the gear ratios for all forward speeds and reverse. (Note: this investigation must be done on a conventional type vehicle.)

(d) How is the accidental engagement of reverse prevented during normal gearchanging?

(e) Considering the position of the gearbox in the vehicle, list the other major parts that would have to be removed before the gearbox itself could be removed.

Project

(i) It has already been stated that most vehicles employ synchromesh devices to simplify gearchanging. Make a study of these devices and describe, with the aid of sketches, how they work.

(ii) Automatic transmissions replace the conventional clutch and gearbox. They make driving much easier, but add about 10% to the price of the car. List the prices of three popular cars with and without automatic transmission and determine whether or not you consider the extra cost is justified.

PROPELLOR SHAFT

In vehicles of conventional layout (front engine—rear wheel drive) the propellor shaft transmits the drive from the gearbox mainshaft to the final drive pinion in the rear axle. To allow for movement of the rear axle relative to the gearbox, the propellor shaft usually has two universal joints and a sliding joint.

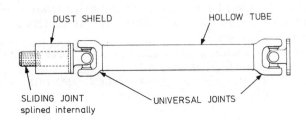

DUST SHIELD   HOLLOW TUBE

SLIDING JOINT
splined internally   UNIVERSAL JOINTS

FINAL DRIVE AND DIFFERENTIAL ASSEMBLY

The pinion is attached to the propellor shaft and drives the crown wheel, which, in turn drives (via the differential) the half shafts which are fixed to the driving wheels. These two gears do therefore transfer the drive through $90°$.

Another function of the final drive gears is to provide a permanent gear reduction in the transmission system.

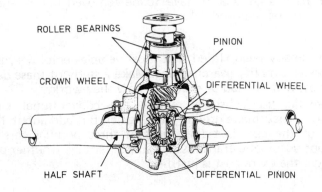

ROLLER BEARINGS   PINION

CROWN WHEEL   DIFFERENTIAL WHEEL

HALF SHAFT   DIFFERENTIAL PINION

DIFFERENTIAL

When rounding a corner, the outside wheel must travel a greater distance, and thus rotate faster, than the inside wheel. Where wheels are not driven, for example, the front wheels of a conventional vehicle, this difference does not matter. However, when the wheels are driving wheels it is important that they can be driven at different speeds and yet at the same time they must both receive equal driving torque. It is the job of the differential to allow this to happen. If it were not done, rapid tyre wear would occur and the handling of the car would be adversely affected.

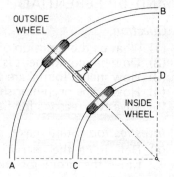

Distance from 'A' to 'B' is greater than distance from 'C' to 'D'.

One disadvantage of a differential can happen when one driving wheel (or both driving wheels) is on a slippery surface such as on ice or in mud. The inherent action of the differential is to increase the speed of the wheel offering least resistance to rotation—at the expense of the speed of the other wheel. Taken to its ultimate if one wheel slips completely, all the speed of rotation goes to that wheel, whilst the other wheel (which may have a firm grip) does not rotate at all. Hence the vehicle will not move!

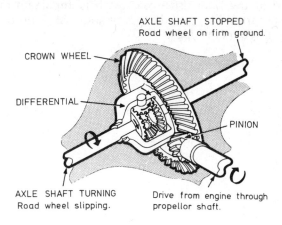

AXLE SHAFT STOPPED
Road wheel on firm ground.

CROWN WHEEL

DIFFERENTIAL

PINION

AXLE SHAFT TURNING
Road wheel slipping.

Drive from engine through propellor shaft.

35

EXERCISE 6C. THE PROPELLOR SHAFT, FINAL DRIVE AND DIFFERENTIAL

Question

(1) What us the function of the differential?
(2) On what component is the differential assembly mounted?
(3) How many 'joints' are there on a propellor shaft?
(4) How is the propellor shaft secured to the final drive pinion?
(5) Give two reasons for fitting final drive gears.

Investigation (State make, model and year of vehicle)

(a) Determine the gear ratio of the final drive by noting how many times the propellor shaft rotates for one turn of *both* the driving wheels.

(b) By tapping the propellor shaft, determine which parts of it are solid and which parts are hollow. Suggest a reason for the hollowness.

(c) With one driven wheel jacked-up* and free to rotate, turn the propellor shaft. Why is it that the wheel on the ground does not move?

(d) Jack-up* both the driving wheels, and rotate them in *opposite* directions *at the same time.* The propellor shaft does not rotate. Why is this?

Project Work

(i) Some racing and high-powered cars fit a 'limited slip' differential. Find out some of the cars that use it and its disadvantages.

(ii) The shape of the final drive gear teeth is important in many ways. It influences their life, quietness, size and cost. Find out all you can about the three main types.

* See 'Jacking-up the Vehicle', page 69.

SECTION 7
THE BRAKING SYSTEM

A moving vehicle must be able to be brought quickly to rest by its brakes. Most cars have two braking systems; one which works on all four wheels and is operated by a foot pedal; and one which operates on the rear wheels only and is operated by a hand lever; this is the parking brake (or handbrake).

When the brakes are applied, pieces of friction material (basically asbestos) are forced against drums or discs which revolve with the road wheels. The friction thus created slows the wheels and eventually stops the vehicle.

HYDRAULIC BRAKING SYSTEM

On the vast majority of modern cars, the brakes are hydraulically operated from the foot pedal and mechanically operated (by rods or cables) from the handbrake lever.

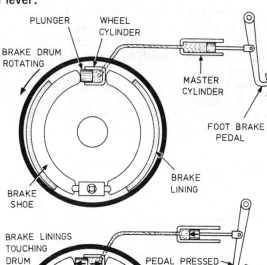

Brake Off

The brake shoes are just clear of the drum allowing it to rotate freely.

PLUNGER

WHEEL CYLINDER

BRAKE DRUM ROTATING

MASTER CYLINDER

FOOT BRAKE PEDAL

BRAKE LINING

BRAKE SHOE

Brake Applied

The transfer of fluid from the master cylinder into the pipeline increases the volume of fluid in the wheel cylinder which forces the plunger and housing apart until the brake linings contact the drum and resist its rotation.

BRAKE LININGS TOUCHING DRUM

PEDAL PRESSED

The wheel cylinder slides and as the plunger moves forward, the cylinder moves backward.

DRUM STOPPED

37

DRUM BRAKES

Complete Hydraulic Braking System

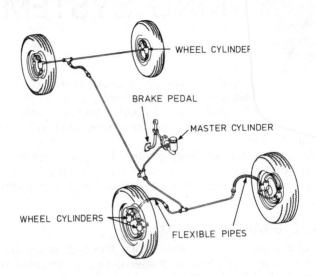

WHEEL CYLINDER

BRAKE PEDAL

MASTER CYLINDER

WHEEL CYLINDERS

FLEXIBLE PIPES

The drawing above shows a hydraulic braking system employing drum brakes at both front and rear. Fluid pressure generated at the master cylinder is applied equally throughout all parts of the system, the brakes are therefore applied with equal pressure to give balanced braking.

The brake 'linings' are either riveted or bonded to the brake shoe, as shown below. As the linings wear down it becomes necessary to replace them, or the brake drums will be damaged owing to metal-to-metal contact, for example, brake rivets scoring the drum.

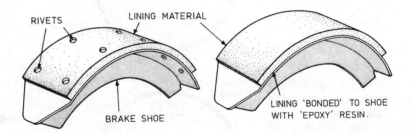

RIVETS

LINING MATERIAL

BRAKE SHOE

LINING 'BONDED' TO SHOE
WITH 'EPOXY' RESIN.

DISC BRAKES

One disadvantage with drum brakes is that severe or prolonged braking can excessively heat-up the brake linings so that their frictional properties are temporarily reduced and the brakes do not work properly. This fall-off in braking performance is known as 'brake fade'.

The heat generated in a disc brake assembly however is more easily dissipated because the rubbing surfaces are exposed to the cooling air and not enclosed inside a drum.

It can be seen from the simplified drawing of the disc brake shown below that the operation is similar in principle to the brake employed on most bicycles.

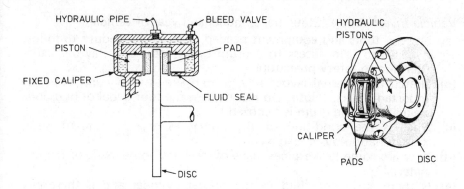

Operation

Friction pads are clamped to the rotating disc by hydraulically operated pistons. It is necessary, as is the case with drum brake linings, to replace the pads after they have become worn.

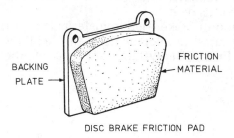

DISC BRAKE FRICTION PAD

EXERCISE 7A. FOOT BRAKES

Questions

(1) Why is it important that all the brakes on a car give balanced braking effort?

(2) How would only one broken brake pipe affect the operation of a hydraulic system?

(3) What is the main advantage of disc brakes when compared with drum brakes?

(4) Why is it advantageous to use hydraulic fluid as a means of transmitting force?

(5) In a hydraulic braking system the master cylinder piston is usually smaller than the wheel cylinder pistons. What is the reason for this?

Vehicle Investigation (State make, model and year of vehicle)

(a) List the tools and equipment needed, and the procedure to follow in order to gain access to a drum brake assembly. Pay particular attention to safety precautions.

(b) Sketch the internal layout of a drum brake assembly and name the important parts. (Note: Do not depress the foot pedal of hydraulic brakes when the drum is removed.)

(c) What is the condition of the linings as regards state of wear, contamination by oil and so on?

(d) Why are some brake pipes made of steel and some made of flexible material?

(e) Is there sufficient fluid in the master cylinder and is there any evidence of brake fluid leaks anywhere on the system? (Note: The system *must never* be topped up with oil.)

Project Work

(i) Both disc and drum brakes are employed in the same system on many modern vehicles. Find out more about this arrangement and give reasons for using it.

(ii) The production of brake friction material is surprisingly interesting. Find out the latest developments in this field. (Note: Ferodo Ltd., Chapel-en-le-Frith, Derbyshire, are one of the world's leading manufacturers.)

HANDBRAKE OPERATION

The handbrake on most cars works on the rear wheels only and is usually mechanically operated. A typical cable arrangement for a handbrake is shown below. One important feature of the system is that both brake units should receive equal pull from the cables. This is achieved by applying the pull from the brake lever through a balancing mechanism which is known as a 'brake compensator'.

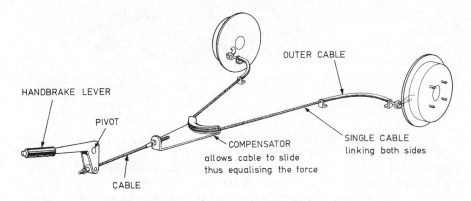

OUTER CABLE

HANDBRAKE LEVER

PIVOT

COMPENSATOR
allows cable to slide
thus equalising the force

SINGLE CABLE
linking both sides

CABLE

When the handbrake is operated, the cables pull on levers which in turn push out the wheel cylinder plungers. The simplified drawing of a rear wheel cylinder incorporating a handbrake mechanism shows how this works.

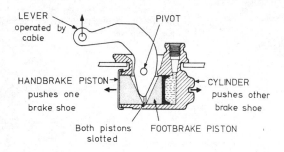

LEVER
operated by
cable

PIVOT

HANDBRAKE PISTON→
pushes one
brake shoe

←CYLINDER
pushes other
brake shoe

Both pistons   FOOTBRAKE PISTON
slotted

Unlike the footbrake, the handbrake must remain locked-on whilst the vehicle is parked. This can be achieved by incorporating a 'ratchet' mechanism at the lower end of the lever. This is the part that clicks as the brake is applied.

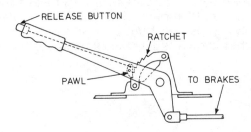

BRAKE ADJUSTMENT

As the friction material wears down, the clearance between the drum and lining, or pad and disc, will increase. It is therefore necessary to provide some form of adjustment in order to keep the clearance to a minimum. With a disc brake assembly, and certain drum brakes, adjustment is done automatically as the brake is operated. However most drum brakes have manual adjusters which should be reset from time to time. Two types of manual brake adjusters are shown below.

Adjusting a brake with a screwdriver through a hole in the drum by turning the large screw (or in some cases a small toothed wheel) which is visible through the hole, clockwise to compensate for lining wear.

Square-headed type of adjuster is rotated clockwise to compensate for lining wear. In this example there are two such adjusters on each front brake.

EXERCISE 7B. HANDBRAKE AND ADJUSTMENT

Questions

(1) What is the purpose of the handbrake 'compensator'?
(2) Why is it necessary to keep the clearance between the lining and drum (or pad and disc) to a minimum?
(3) What effect will the presence of air in the hydraulic system have on the operation of the footbrake?
(4) Why is the handbrake mechanism mechanical rather than hydraulic?
(5) What are the dangers of relying on leaving the car 'in gear' rather than using the handbrake?

Vehicle Investigation (State make, model and year of vehicle)

(a) Make a sketch to show the arrangement of the handbrake mechanism from the lever to the rear wheels; show clearly the compensator.
(b) Examine the handbrake mechanism for wear and check for satisfactory operation; list any faults you find.
(c) Examine a front and rear brake assembly and make sketches to show the type of brake adjusters employed.
(d) Take the proper safety precautions,* then, after consulting the car handbook, raise each wheel in turn clear of the ground and adjust the brakes as necessary.
(e) What form of adjustment, if any, is provided in the handbrake mechanism?

Project Work

(i) Some vehicles employ what is known as a transmission handbrake (for example, Bedford and Land Rover). Find out more about these brakes and sketch and describe one type. State the make and model of vehicle using it.
(ii) 'Servo units' are employed in many modern (especially fast, high-powered) vehicles. Find out more about these units, how they work and why they are fitted.

* See 'Jacking-up the Vehicle', page 69.

SECTION 8
STEERING

So far as the driver is concerned, the precision and 'feel' of the steering is of considerable importance. Movement of the steering wheel should cause an accurate response of the road wheels, which should not be deflected from their correct paths by body roll or bumps in the road surface.

The rotational movement of the steering wheel must be changed into the lateral (side to side) movement needed to turn the road wheels in the desired direction. This is done by the gearing at the lower end of the steering column which also multiplies the driver's effort and so makes it easier for him to turn the wheels.

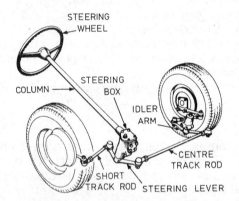

THREE-PIECE TRACK ROD & STEERING BOX

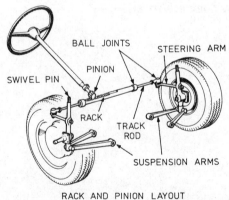

RACK AND PINION LAYOUT

44

STEERING GEOMETRY

To ensure true rolling of the road wheels when a car is travelling in a straight line is easy. Quite simply, all the wheels must point straight ahead. However, on rounding a bend, all the wheels should roll in arcs whose centres are at the centre of the bend. Centre-pivoted farm carts and even children's 'bogies' or 'trucks' achieve this very well.

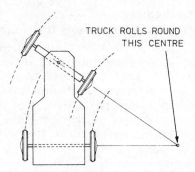

Centre-pivoted axle. All wheels follow different arcs but have a common centre.

On all but some of the earliest cars, the requirements of a common turning centre for all the wheels is satisfied by mounting each front wheel on a short 'stub axle', each having its own pivot or 'king pin'. The angle through which each front wheel is turned can differ as required.

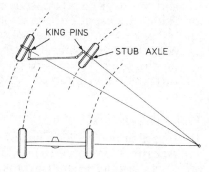

As can be seen above, when rounding a bend the inner front wheel must turn through a greater angle than the outer front wheel. The difference between these two angles is kept correct by the 'Ackerman linkage', which means no more than inclining the steering arms to achieve the simple geometrical layout shown.

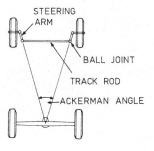

The basic principle is not affected with the more complicated steering layouts used with independent front suspension.

45

EXERCISE 8. STEERING

Questions

(1) What unit changes the rotary movement of the steering wheel into the lateral movement needed by the track rod (or rods)?
(2) Whereabouts are ball joints fitted to the steering mechanism?
(3) Why would the type of steering shown at the top of the previous page not be suitable for a car?
(4) In order to achieve true rolling when rounding a bend, what conditions must the geometry of the road wheels satisfy?
(5) Which wheel turns through the greatest angle when rounding a bend?

Vehicle Investigation (State make, model and year of vehicle)

(a) How many turns of the steering wheel are required from maximum left to maximum right lock?
(b) What is the ratio of movement (the gear ratio) of the steering wheel and the steering box output shaft?
(c) Sketch the layout of the steering mechanism and name the parts.
(d) Show by a sketch how the length of a track rod may be adjusted.
(e) With the car on a clean, level surface, and with the aid of chalk lines on the floor, attempt to check that when on lock all the wheels do turn round a common centre. (See middle drawing on previous page.)

Project Work

(i) Investigate the origins of 'Ackerman steering'. (Dr Rudolph Ackerman patented the system but did not invent it.)
(ii) Cars are only steered by their front wheels. Find out why it is that all four wheels, or the rear wheels only are not used for this purpose. (Note: Stacker trucks do use rear wheel steering.)

SECTION 9
THE ELECTRICAL SYSTEM

SAFETY

There is no possibility of receiving an electrical shock from the normal 12 V motor car electrical system. The only exception is the high tension ignition system, and, if for example you touched the bare end of a spark plug lead whilst the engine was running, it would probably cause you to jump. However, beware of sparks or overloaded cables which could cause a fire.

It is essential, when dealing with batteries, to avoid any contact with acid and to ensure that sparks or naked flames cannot occur in the vicinity of the battery top. This is to prevent battery gases, which are given off during charging, from exploding.

MAIN ELECTRICAL COMPONENTS

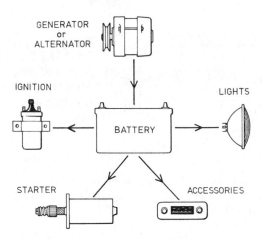

The battery acts as a reservoir of electrical power and is the heart of the car's electrical system. All the other electrical components take current from the battery; that is except the generator (or alternator) which charges the battery whilst the engine is running.

ELECTRICAL CIRCUITS

The Need for a Complete Circuit

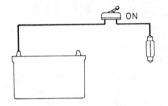

Although a bulb may be connected to a battery as shown it will not light.

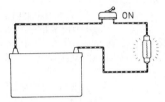

The current must be able to go to the bulb (or other item of electrical equipment) pass through it, and return to the battery. That is, a complete circuit is needed.

Earth Return

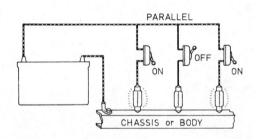

These bulbs are wired *in PARALLEL*. Each may be operated independently of the other. Most car electrical components are so connected.

By earthing one side of the battery to the metal body or chassis and earthing each electrical component (usually simply by bolting it in position), the number of cables needed is halved.

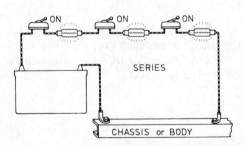

Current flow is weak since it has to pass through all the resistances (bulbs). Interruption of current flow at *any* point puts out all the bulbs. These bulbs are wired *in SERIES*.

VEHICLE LIGHTING

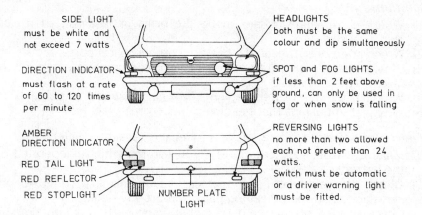

SIDE LIGHT
must be white and
not exceed 7 watts

DIRECTION INDICATOR
must flash at a rate
of 60 to 120 times
per minute

AMBER
DIRECTION INDICATOR

RED TAIL LIGHT

RED REFLECTOR

RED STOPLIGHT

NUMBER PLATE
LIGHT

HEADLIGHTS
both must be the same
colour and dip simultaneously

SPOT and FOG LIGHTS
if less than 2 feet above
ground, can only be used in
fog or when snow is falling

REVERSING LIGHTS
no more than two allowed
each not greater than 24
watts.
Switch must be automatic
or a driver warning light
must be fitted.

The law relating to the lighting of motor vehicles is extremely complicated, but some of the major points are shown above. Apart from checking that all the lights are kept clean and in good working order, probably the next most important feature is to check that the headlamps are correctly aimed.

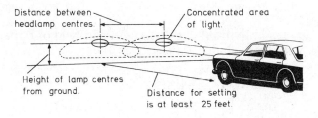

Distance between headlamp centres.

Concentrated area of light.

Height of lamp centres from ground.

Distance for setting is at least 25 feet.

49

EXERCISE 9A. ELECTRICAL

Questions

(1) Which component is said to be the heart or reservoir of a car's electrical system?

(2) Which component feeds the battery with electrical current?

(3) Why are nearly all automobile electrical components wired on the 'earth return' system?

(4) Why is it that almost all the lights on a car are wired 'in parallel'?

(5) Why is it important that headlamps are correctly aimed?

Vehicle Investigation (State, make, model and year of vehicle)

(a) Make a list of every component you can see which is operated by the battery.

(b) Check the alignment of the car headlamps according to the instructions on the preceding page. Report any defect.

(c) Check all the lights to see if they operate properly. List any defects.

(d) By tracing the thickest cables on the car (starting from the battery) decide which component requires the most current.

(e) With cable, bulbs and a battery, connect them up with the bulbs in series and then in parallel. Observe the difference in illumination.

Project Work

(i) In many Continental countries cars must use yellow headlamps. Which countries are these and what are the advantages of this practice?

(ii) Many cars are now fitted with alternators instead of dynamos. How do they work and what are their advantages?

BATTERIES

Motor car batteries are of the secondary-cell, lead-acid type; thus each of the six cells that together make up a 12-V battery can be charged, discharged and recharged repeatedly. The action is 'reversible'. This is unlike, for example, a transistor radio battery which is of the primary-cell (non-reversible) type and must be discarded when discharged.

Passing current into a secondary cell causes chemical changes to occur within the cell. Taking current from the cell causes these changes to be reversed.

Basically a lead-acid, secondary-cell consists of lead plates, separated from each other, and immersed in dilute sulphuric acid.

Cell about to receive initial charge of electricity.

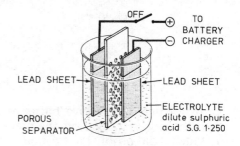

Cell being charged. Bubbling and the change of colour of the positive plate are obvious indications of chemical action.

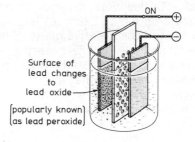

The source of charge is removed after a few minutes and the cell is capable of lighting a small bulb.

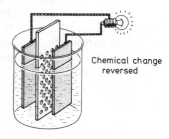

Lead-Acid Car Battery

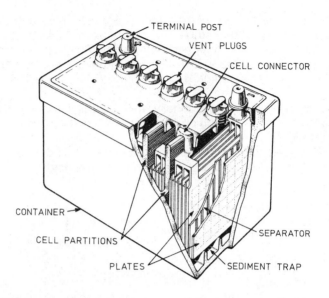

In practice automotive batteries are much more refined and robust than the cell described on the previous page. The outer container made of tough, acid-proof, hard rubber or plastic is divided up internally into cell compartments. Each plate is made from antimonial-lead, cast in the form of a lattice-work or grid. The spaces in the grid are filled with active material which becomes lead peroxide in the positive plates and spongy lead in the negative plates. Several plates are used in each cell so as to increase the storage capacity. A top cover prevents spillage and contamination, but is fitted with a cap (or caps) to allow distilled water to be added to replace that lost by evaporation.

Negative and positive terminal pillars are provided on which to mount the battery leads.

Battery Maintenance
(b) The electrolyte must be kept topped up with distilled water to about $\frac{1}{8}$ inch above the top of the plates.
(b) The terminals must be kept clean and tight.
(c) The battery must be held firmly in position.
(d) The top of the battery should always be wiped clean and dry.
(e) The terminals and fixing bolts can be prevented from corroding by smearing lightly with petroleum jelly.

CABLES

Modern cars contain an enormous number of electrical cables, which can appear bewildering at first sight. However, on closer inspection and acquaintance, it can be realised that they do conform to a pattern and that individual cables can be recognised. When their function is not obvious, cables are mainly identified by their colour. Thus a red cable, say under the bonnet, may easily be traced behind the dash panel.

Very often a cable has two or more colours, usually with one predominating; the other less obvious colour is known as the 'tracer'. The main colour indicates the basic circuit and the tracer colour the auxiliary circuit to which the cable belongs. For example, the lead to a headlamp main beam filament may be blue only, while the lead to the dipped filament may have a basic colour of blue with a tracer colour of red.

FUSES

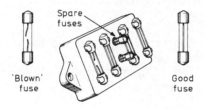

FUSE BLOCK ASSEMBLY

A fuse is a useful safety device, placed in a circuit to protect the component and cables of that circuit from damage that may result from electrical overload. It also prevents the protected circuit from possibly starting a fire.

Essentially the fuse is a short length of low melting point wire. If too much current flows through it, it overheats, melts and thus stops any further current flow.

EXERCISE 9B. BATTERIES and WIRING

Questions

(1) Why are vehicle batteries of the secondary-cell type?
(2) How many cells are there in 6 V, 12 V and 24 V batteries respectively?
(3) What is the basic constituent of battery plates and what is the function of the separators?
(4) Of what is battery electrolyte composed and with what should it be topped-up if necessary? State the precautions necessary when handling electrolyte.
(5) How can individual cables in a car wiring system be identified?

Vehicle Investigation (State make, model and year of vehicle)

(a) Examine the battery and determine what maintenance it requires.
(b) List five of the major electrical circuits in the car and the colour of the cables used to identify them.
(c) Find the fuses within the car and determine which circuits they protect. Make a sketch of the fuse block and label each fuse with the circuits it protects.
(d) Sketch three of the many types of electrical cable connectors or terminals found on a car.
(e) Find out which circuits operate only when the ignition is switched on. List these and explain why they are arranged in this way.

Project Work

(i) In many instances nowadays, current is conveyed via 'printed circuits' instead of the conventional cables. Find out more about these, where they are used and their advantages and disadvantages.
(ii) It is often necessary to fit extra lights and other electrical accessories on cars. List the accessories that you consider desirable, find their approximate price and determine how they should be connected into the electrical system.

SECTION 10
TYRES AND WHEELS

Tyres have two main func-
tions; the obvious one of
providing cushioning properties against road irregulari-
ties; and to control the
path of the vehicle. This
last-mentioned point is
vital. For example, the only
contact a car has with the
road is four small areas of
tyre tread each little larger
than the flat of your hand.
It is only via these contact
areas that the car's position
on the road can be controlled, and the design and condition of the tyres has an
enormous effect on the
handling stability and safety
of a vehicle.

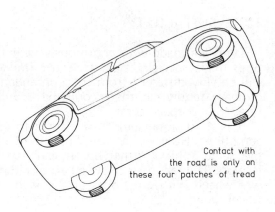

Contact with
the road is only on
these four 'patches' of tread

Tyres are of a much more complex construction than their external
appearance suggests. Hidden from view under the familiar black rubber
exterior is a scientifically designed casing (or framework) made up from
fabric such as nylon or rayon, together with steel wire, and in some
cases what is virtually steel fabric. The two main types of tyre take
their name from the form of casing construction; radial-ply and
cross-ply.

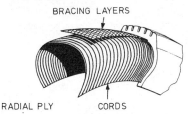

BRACING LAYERS

RADIAL PLY   CORDS

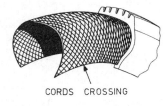

CORDS CROSSING

CROSS PLY

The fabric cords are disposed
radially (hence radial-ply), but
with the addition of bracing
layers of fabric or steel mesh
under the tread area.

Layers (or plies) of fabric are
arranged so that the cords of the
various plies cross each other at
an angle of about 45°.

RADIAL and CROSS-PLY TYRES—RELATIVE MERITS

At present, the radial-ply tyre is becoming more and more popular and at least one European company no longer manufactures any other type. This is because its advantages, such as longer tread life and better road holding and its slightly lower rolling resistance (which marginally improves m.p.g.), outweigh its disadvantages of higher cost, greater transmission of road thump and heavier low-speed steering effort.

TYRE SUITABILITY

Tyres are made with enormously varied specifications. For example, tyres made specially to suit electric (milk-float type) delivery vehicles will be very different from those intended for racing cars. Even for an ordinary family car there could be a wide choice; such as from snow tyres to high-speed tyres.

It is most important to use tyres to suit the car and its operating conditions. However, it is *vital* not to 'mix' radial and cross-ply tyres on the same vehicle, otherwise it can handle in a most unpredictable fashion. The law does allow a *pair* of radial-ply tyres to be fitted to the *rear* wheels with cross-plies at the front, but *not* radials on the front wheels only.

RADIAL-PLY TYRE CONSTRUCTION

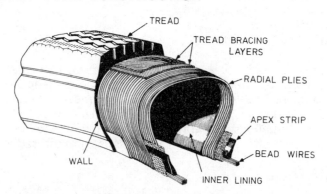

RADIAL-PLY TYRE CONSTRUCTION

WHEELS

Pressed steel wheels. These are the type widely used on ordinary family cars. The centre disc and the rim are welded together to form one strong unit. Very often the wheel is ornamented by a bright metal wheel disc.

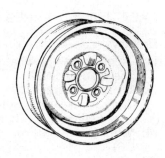

Light alloy sports wheels. It is a help towards better road holding if the wheels can be made as light as possible. However, they must also be strong and this necessitates the use of more expensive metals, for example magnesium alloy. Appearance is good.

Rostyle wheels. These too are of pressed steel, but have a chunky, sporty appearance and require little or no embellishment.

Wire-spoked wheels. In one form or another this type of wheel has been in continuous use since the time of the earliest motor cars. They have a sporty appearance, possess some resilience and allow reasonable air flow to the brakes. In time the spokes can work loose in service. They are awkward to clean and are unsuitable for tubeless tyres.

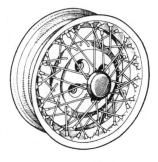

EXERCISE 10   TYRES and WHEELS

Questions

(1) Approximately what area of an average car is in direct contact with the road surface?
(2) What is the main difference between radial and cross-ply tyres?
(3) If a car owner has only two radial-ply tyres, whereabouts on the car may he legally fit them?
(4) What is the purpose of the tread grooves (see the diagram on page 56)?
(5) Why is it important not to use a vehicle with worn tyres?

Vehicle Investigation (State make, model and year of vehicle)

(a) Inspect the tyres and estimate the percentage wear that has taken place. Tabulate the results.
(b) Inspect the tyres for any defects such as uneven wear, bulging or cut side walls, and so on. Measure (in millimetres) the depth of their treads. Report on your findings.
(c) Test the tyre pressures (including the spare tyre) of several vehicles and check the results against the recommended pressures.
(d) Examine the wheels of several cars, note any defects you see and sketch the design you most like.
(e) See if you can determine any external features that distinguish radial and cross-ply tyres.

Project Work

(i) Find out how a tyre is made.
(ii) Illustrate the history of the tyre from J. B. Dunlop's original to those of today. (Note: The Education Section of the Dunlop Co. Ltd, 25 St James's St, S.W.1, are very helpful.)

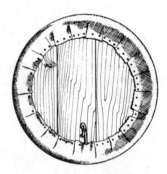

A replica of John Boyd Dunlop's experimental tyre.

SECTION 11
THE SUSPENSION

The suspension system helps the vehicle to travel over uneven surfaces without transmitting road shocks to the occupants.

One of the simplest forms of suspension is the leaf spring and beam axle arrangement. This is employed on most commercial vehicles and at the rear of many cars. The leaf spring is made up of a number of flexible steel strips.

BEAM AXLE ARRANGEMENT EMPLOYING LEAF SPRINGS

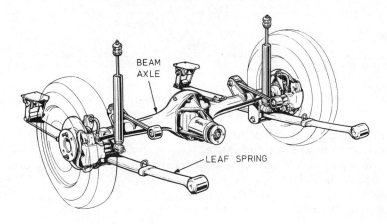

BEAM AXLE

LEAF SPRING

ACTION OF A LEAF SPRING

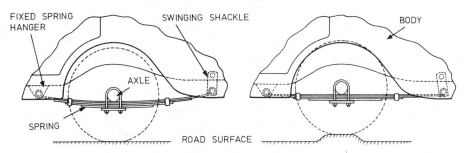

FIXED SPRING HANGER

SWINGING SHACKLE

BODY

AXLE

SPRING

ROAD SURFACE

The drawing above illustrates the action of a semi-elliptic type of leaf spring. As the wheel rides over a bump, the spring flexes as shown and allows the wheel to rise and fall without unduly lifting the body of the vehicle.

59

INDEPENDENT SUSPENSION

With the beam axle arrangement, owing to the fact that the road wheels are on a single axle, when one wheel rises over a bump the axle tilts and the other wheel is affected. For this and other reasons most modern cars are 'independently' sprung.

The action of independently sprung wheels compared to the beam axle arrangement is shown below.

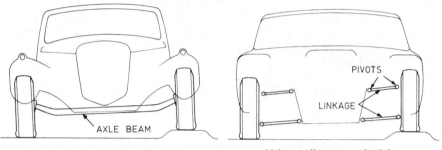

Movement of axle tilts both wheels.

Linkage allows one wheel to rise and fall independently.

Some advantages of independent suspension when compared with the beam axle arrangement are:

(1) Improved passenger comfort.
(2) The vehicle will 'corner' better.
(3) The tyres are kept in better contact with the road and this improves road holding.
(4) Allows a greater wheel deflection.
(5) Allows the engine to be positioned lower in the frame.

The main disadvantages are that the tyres tend to wear unevenly and the suspension is more complicated and costly.

COIL SPRINGS

One type of spring used on many independent suspension layouts is the coil or helical spring. Coil springs will allow a greater wheel deflection and can be made to give a softer ride than the leaf spring.

An independent front suspension arrangement employing a coil spring is shown below.

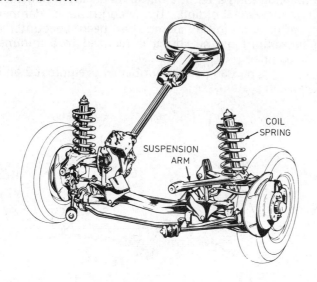

INDEPENDENT REAR SUSPENSION

To improve the ride and handling of the vehicle in general, many modern vehicles are fitted with independent suspension at the rear. When this is done, the final drive housing is usually bolted to the frame and as can be seen below short drive shafts fitted with universal joints transmit the drive to the wheels.

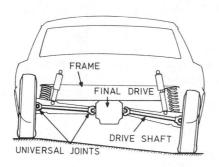

DAMPERS (shock absorbers)

Owing to the flexibility of a spring, once the spring has been deflected as a result of say the wheel rising over a bump, it will continue to flex (that is, vibrate or oscillate) for a period of time. This vibrating action, if unchecked, would tend to allow the car body to continue bouncing long after the bump, and have a detrimental effect on stability and roadholding. By incorporating dampers in the suspension, after the initial bump has been encountered, further oscillation (vibration) of the spring is reduced to a minimum by the action of the dampers.

The telescopic-type damper shown below is employed on very many vehicle suspension systems.

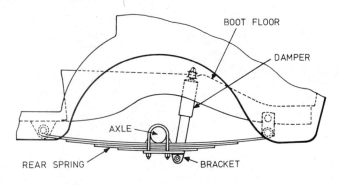

Principle of Operation (Telescopic type)

As the damper is compressed or extended, fluid is forced from one side of the piston to the other through small drillings. The limited rate at which the fluid can flow creates a resistance to movement which prevents undue flexing of the road spring. The damper normally incorporates a 'two way' valve and reservoir to allow for the volume of piston rod entering and leaving the cylinder but these have been ignored for the sake of simplicity.

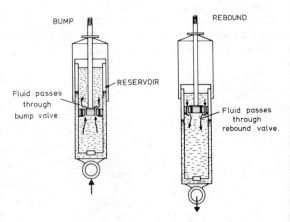

EXERCISE 11   SUSPENSION

Questions

(1) What part of the vehicle, other than the actual suspension system, will absorb road shocks?

(2) Why is the leaf spring attached to a 'swinging shackle' at one end?

(3) One feature of the leaf spring is its inherent friction and consequent 'self damping' (or shock absorber) action; consider the action of a leaf spring and explain how this is achieved.

(4) Why is independent suspension employed more commonly at the front of vehicles than at the rear?

(5) What would be the symptoms of worn dampers on a vehicle?

Vehicle Investigation

Examine a vehicle which has independent suspension at the front and leaf springs at the rear. (State make, model and year of vehicle)

(a) Make a simple sketch to show how the leaf springs are attached to the axle.

(b) How many leaves has each spring? Make a list of any signs of wear, breakage or other defects.

(c) Make a simple sketch to show the arrangement of the front suspension (one side of the vehicle only).

(d) Which of the two springs, front or rear, would be easier to change in the event of a failure and why?

(e) What limits the maximum deflection of (i) the front springs (ii) the rear springs?

Project Work

(i) There are a number of forms of springing, other than leaf and coil, used on vehicle suspension systems. Find out what these are, on what vehicles they are used and state some of their advantages. (One example: rubber springs on 'Minis'.)

(ii) Spring dampers are rather more complex in practice than their simple principles might suggest. Discover exactly how one type of modern damper works. (Note: Armstrong Patents Co Ltd, of Melton, Yorks, and Girling Ltd, Tyseley, Birmingham, are well known makers of these components.)

SECTION 12
MAINTENANCE

If a vehicle is kept in a sound and roadworthy condition, the number of breakdowns and emergency repairs will be kept to a minimum; it must therefore be subjected to a system of regular routine maintenance.

This will include the following tasks:

(a) Routine servicing: oil changing, chassis lubrication, minor mechanical and electrical maintenance.
(b) Cleaning and valeting: washing and polishing the car, cleaning the interior, cleaning the underbody.
(c) Body maintenance: minor body repairs, care of door locks and hinges, care of paintwork.
(d) Periodic checks: 'weekly maintenance' as mentioned below; plus less frequent, but more searching checks of items such as condition of tyre side-walls, inspection of underbody for corrosion and ensuring that at all times the vehicle is well above the minimum standard required to pass the Ministry Test.

WEEKLY MAINTENANCE

It is normal to carry out the following tasks every week or before departing on a long journey:

(1) Check engine oil level.
(2) Check water level in radiator.
(3) Test tyre pressures (including spare).
(4) Check electrolyte level in battery.
(5) Ensure that all lights are in working order.
(6) Check contents of screen-washer reservoir.

"I didn't know it needed oil as well as petrol"

ROUTINE SERVICING

Vehicle servicing is carried out at intervals which are determined either by time or mileage; for example, a vehicle doing average mileage (about 1 000 miles per month) would normally be serviced on a mileage basis. A vehicle doing only a small mileage may be serviced on a time basis, such as every three months.

It must be stressed that precise servicing requirements can vary considerably from model to model and the vehicle manufacturer's instructions should be strictly adhered to. The routine servicing suggestions contained herein, are intended only as a general guide in respect of a conventional vehicle, and with the emphasis on what can be done without the need of specialised knowledge or equipment.

Routine servicing is often split into two categories:

(a) Minor servicing—usually every 5 000/6 000 miles.
(b) Major servicing—usually every 10 000/12 000 miles.

The two servicings are distinguished by the extent of the work carried out. A typical minor service would include the following operations:

Drain engine oil and refill with recommended lubricant.
Fit new engine oil filter element.
Lubricate all lubrication points: handbrake cables, steering ball joints, etc. (Note: Some modern vehicles have no greasing points.)
Check brake and clutch reservoir fluid levels.
Check brake pedal travel and adjust brakes if necessary.
Check fan belt tension and adjust if necessary.
Check water level in engine cooling system, battery electrolyte level and tyre pressures.
Make visual inspection of brake pipes and hoses.
Check for oil leaks on engine and transmission.

A major service would include all the items stated above but with the addition of a number of other operations which need not be carried out so frequently.

The additional operations carried out on a typical major service would normally be:

Engine

Check valve rocker (tappet) clearances and adjust as necessary. Clean or replace contact points and reset gap.

Clean sparking plugs or replace and reset gaps.

Check ignition timing.

Check free play at clutch pedal and adjust if necessary.

Lubricate dynamo rear bearing.

Lubricate water pump.

Lubricate distributor as necessary.

Clean engine breather.

Renew air cleaner element.

Transmission

Drain and refill gearbox and rear axle.

Brakes

Inspect brake linings (or pads) for wear and blow the dust from linings and drums.

Suspension and Steering

Check nuts and bolts on suspension and steering for tightness.

Check moving parts on suspension and steering for wear.

Check 'wheel alignment'.

Check oil level in steering box.

Lights

Check headlamp alignment and adjust if necessary.

The various settings (such as ignition and tappets), capacities, recommended lubricants, lubrication points and the like can be obtained from the vehicle handbook.

Typical settings are:

Sparking plug gap	0·025 in (0·65 mm)
Contact breaker gap	0·015 in (0·40 mm)
Valve clearance (hot)	
Inlet	0·010 in (0·25 mm)
Exhaust	0·012 in (0·30 mm)

A popular method of presenting lubrication data is shown opposite.

Lubrication Chart and Recommended Lubricants

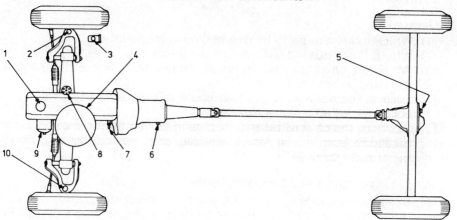

Lubrication Chart Key:

(1) Engine oil filter.
(2) Suspension ball joint.
(3) Brake fluid reservoir.
(4) Air cleaner element.
(5) Rear axle oil level plug.

(6) Gearbox oil level plug.
(7) Engine oil dipstick.
(8) Distributor.
(9) Engine oil filter element.
(10) Suspension ball joint.

RECOMMENDED BRANDED LUBRICANTS—UNITED KINGDOM

Manufacturers	Engine	Gearbox	Rear axle	Lubricant gun
Petrofina	Fina Supergrade Motor Oil 20W/50	Fina Pontonic XP SAE 90/140	Fina Pontonic XP SAE 90/140	Fina Marson LM 2
Shell	Shell Super Motor Oil 100	Shell Spirax 90 EP	Shell Spirax 90 EP	Retinax AM
Texaco	Havoline Oil 10W/30	Multigear Lubricant 90	Universal Thuban 90	Molytex Grease 2
BP	Super Visco-Static SAE 20W/50	BP Gear Oil SAE 90 EP	BP Gear Oil SAE 90 EP	Energrease L21M
Castrol	Castrol GTX	Castrol Hypoy	Castrol Thio-Hypoy FD or Hypoy	Castrol MS.3 Grease
Duckhams	Duckhams Q20/50 Motor Oil	Duckhams Hypoid 90	Duckhams Hypoid 90	L.B.M. 10 Grease
Esso	Uniflo	Esso Gear Oil GX 90/140	Esso Gear Oil GX 90/140	Esso MP Grease (Moly)
Gulf	Gulfpride Single G 10W/30	Multi-purpose Gear Oil 90	Multi-purpose Gear Oil 90	Gulflex Moly
Mobil	Mobiloil Super	Mobilube GX 90	Mobilube GX 90	Mobilgrease Super

Note: Castrol Ltd, Marylebone Road, London NW1 5AA, publish free lubrication charts.

EXERCISE 12A. ROUTINE SERVICING

Questions

(1) Which items require to be checked most frequently?
(2) Why is it essential to change engine oil regularly?
(3) At about what mileage intervals should (a) minor and (b) major servicing be required?
(4) Suggest the possible faults that could arise as a result of failing to check water, oil and electrolyte levels.
(5) Apart from the obvious benefits, such as minimising the chance of breakdowns from regular vehicle servicing, what other additional benefits might there be?

Vehicle Investigation (State make, model and year of vehicle)

(a) Examine a vehicle and with the aid of simple sketches describe how the correct oil level for the engine, gearbox and rear axle can be determined.
(b) State the number and location of the lubrication points on the vehicle.
(c) Check the following and make out an appropriate report:

    (1) oil levels;
    (2) cooling water, screen washer reservoir and battery electrolyte levels;
    (3) brake and clutch master cylinder levels;
    (4) tyre pressures;
    (5) engine and transmission for oil leaks.

(d) Examine the vehicle handbook and list the items to receive attention during major servicing that are in excess of those dealt with during the minor service.
(e) List any points where you consider that corrosion has seriously weakened the structure of the vehicle.

Project Work

(i) Modern engine oils may contain up to 25% of substances known as additives which improve the oil and give it properties it would not otherwise possess. What are the principal additives and how do they work?
(ii) Estimate how much it would cost per year to service a vehicle (labour and materials), assuming an annual mileage of 12 000.

JACKING-UP THE VEHICLE

From time to time it becomes necessary to raise one or two of the vehicle's wheels off the floor, for example to adjust the brakes or change a wheel. When carrying out this operation a safe working procedure must be adopted.

(1) Ensure that the vehicle is on firm, level standing (for example, concrete).
(2) Chock at least one wheel which is to remain on the ground, so that it can roll neither forward nor backward.
(3) Ensure that the jack is in good condition.
(4) Place the jack under a strong solid part of the vehicle, such as under the axle or cross-member, ensuring that no pipes or cables etc. are trapped. (Note: On older cars it can be very unsafe to trust the built-in jacking points.)
(5) Raise the vehicle and at the same time keep a continual check on the position of the jack.
(6) Support the vehicle using 'axle stands' or suitable blocks. *Never* work on a vehicle which is supported only by a jack.

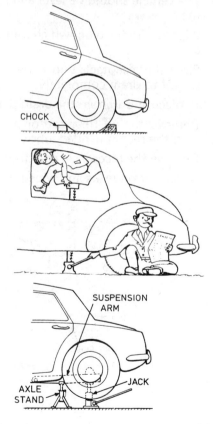

Chock the wheels remaining on the ground.

Jack-up under a strong part of vehicle.

Use axle stands (or wooden blocks) to give additional safe support.

69

OIL CHANGING

When draining the engine, gearbox or rear axle oil, the equipment required and procedure is as follows:

Equipment

Suitable good-fitting spanners to suit drain plugs.

A receptacle large enough to receive the oil.

Sufficient new oil of correct type and grade.

Hand wipers.

Clean oil-measuring can and pourer.

(Note: Some rear axles and gearboxes have awkwardly-positioned filler plugs. Plastic squeeze-bottle fillers, with long flexible spouts, easily overcome this problem.)

Procedure

The vehicle should be level and preferably have just been used so that the oil is warm.

Wipe the drain plug, which will be situated at the base of each unit, and slacken.

Place the receptacle under the drain plug, remove the plugs and allow the oil to drain completely.

Replace and tighten the drain plug.

Remove the oil filler cap or plug, refill the unit to the correct level with the recommended lubricant and replace the cap or plug.

Remove the receptacle and dispose of waste oil.

Check for oil leaks and wipe up any oil which may have been spilt.

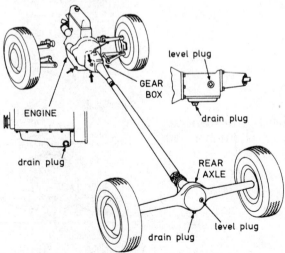

VALVE CLEARANCE

It is important to maintain a clearance between the valve-operating mechanism and the valve itself. This being due to the fact that as the valve temperature increases, its length also increases owing to expansion. If therefore the clearance was insufficient, when the engine was hot the valve mechanism could hold the valve off its seat; this would create loss of compression and cause loss of power and burning of the valve seat. The illustration below shows where the valve clearance can be checked and where the adjustment is made on an o.h.v. engine.

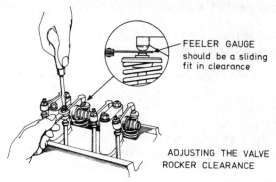

FEELER GAUGE
should be a sliding
fit in clearance

ADJUSTING THE VALVE
ROCKER CLEARANCE

Procedure for Checking Valve Clearance

(1) Rotate the engine until the valve to be set is fully open. (On the vast majority of engines the valve appears to be down.)

(2) Rotate the engine through exactly one revolution and then check the valve clearance. Repeat this procedure for each valve in turn.

FAN BELT TENSION

The method of checking the fan belt tension and a typical adjusting arrangement is shown below.

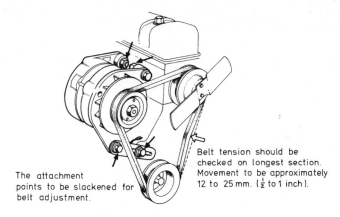

The attachment
points to be slackened for
belt adjustment.

Belt tension should be
checked on longest section.
Movement to be approximately
12 to 25 mm. ($\frac{1}{2}$ to 1 inch).

EXERCISE 12B. JACKING, OIL CHANGING, TAPPETS and FAN BELT

Questions

(1) Why is it necessary to have the vehicle, and especially the jack itself, on a firm base when jacking-up?

(2) Why is it important not to work under a vehicle supported only by a jack?

(3) How can an engine that has no starting handle be rotated for an operation such as setting tappets?

(4) What components are likely to operate inefficiently if the fan belt is too slack?

(5) Why is it essential when a car is jacked-up to chock the wheels remaining on the ground?

Vehicle Investigation (State make, model and year of vehicle)

(a) Check the condition of the fan belt by looking for signs of cracking or fraying; test its tension and look to see how it can be adjusted.

(b) Find out from the vehicle handbook the correct tappet clearances. Check all the clearances and list any discrepancies.

(c) Look under a vehicle, make a 'worm's eye' sketch of the layout and indicate on the sketch the places where you consider it is safe to lift up by using a jack.

(d) Inspect the 'built-in' jacking points of the vehicle and report on their condition.

(e) List the quantity, grade and type of oil needed for the engine, gearbox and rear axle.

Project Work

(i) To always have the engine fan rotating is very wasteful of power. It is now possible to buy thermostatically controlled, electrically operated fans. Find out how they work and more about their advantages and disadvantages.

(ii) Belt drive is not only used for fans and dynamos, it is also employed for transmission purposes and to drive camshafts. Find out more about these special belts. (Note: Daf cars have belt drive.)

ROADWORTHINESS TESTING

Four of the areas covered by the Ministry test regulations are: steering, brakes, lights and tyres.

The items checked under these headings are:

Steering

steering box,
steering linkage,
king pins and bushes or stub-axle ball joints,
suspension pivots,
road springs and dampers,
mounting area on chassis frame for steering gear and suspension.
wheel bearings.

Brakes

efficiency of footbrake—minimum 50%
efficiency of handbrake—minimum 25%
condition of both flexible and steel brake pipes,
hydraulic system for leaks (wheel cylinders, etc.),
footbrake pedal travel,
free play in handbrake,
handbrake cables and linkages,
chassis and road springs.

Lights

operation of all lights,
position of lights on vehicle,
condition of rear reflectors,
headlamp alignment,
condition of wiring and security of lights on bodywork.

Tyres

tread depth (the tread must have a depth of at least 1 mm for the full circumference of the tyre and cover at least 75% of the tread width),
cuts or wear on the tyre sidewall,
correct location of tyres if 'cross-ply' and 'radial ply' are employed on one vehicle.

EXERCISE 12C. VEHICLE SAFETY CHECK

Vehicle Investigation

Carry out the following vehicle examination and make out a report to ascertain its condition. List any items that you consider need attention. (State make, model and year of vehicle.)

Steering

Check: the free play at the steering wheel,
the steering linkage ball joints,
that the steering box or rack assembly is securely bolted in position,
the king pins or stub-axle ball joints and the suspension pivots for
  wear.

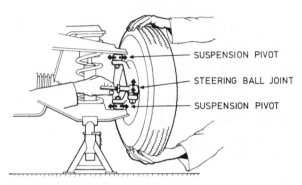

SUSPENSION PIVOT

STEERING BALL JOINT

SUSPENSION PIVOT

Brakes

It is usual to drive the vehicle and use a 'brake efficiency meter' for this test. However if this is not possible, check the footbrake pedal travel, the free play in the handbrake and examine the brake pipes and hoses.

Lights

Check: that all the lights are working and securely mounted on the
  body,
that the rear reflectors are not faded,
the headlamp alignment,
that the wiring to the lights is in good condition and properly clipped
  in position.

Tyres

measure the tyre tread depth (a tread depth gauge is helpful but not
  vital),
check the tyres for cuts, bulges and abnormal wear.

In addition to the checks already carried out, the condition, particularly from the corrosion aspects of the body, chassis frame and seating should also be taken into account in the report.

SECTION 13
CLEANING AND VALETING

WASHING

Regular care of the car bodywork is necessary if the appearance and general condition are not to deteriorate. To protect the car exterior against the action of rain, dirt and air pollution, the bodywork should be washed frequently using a soft sponge and plenty of water containing a mild detergent. Care must be taken not to scratch the bodywork by rubbing mud or grit into the surface, hence the reason for using plenty of water.

It is advisable, when washing a car, to begin at the bottom and work upwards, this prevents the formation of stains (which are difficult to remove) caused by water running through the dirt on the body. Another advantage of working upwards is that it is easier to see which areas have not yet been washed.

After washing, the surface of the car should be dried-off with a clean, damp chamois leather or sponge which itself must be frequently rinsed. Washing can be speeded up by using a specially designed brush, containing soap pellets in the hollow handle, which itself is adapted to fit on to a hosepipe. When the water is turned on, soapy water is continually forced through the brush head. The pellets can be removed to swill the car.

CARE OF BRIGHT TRIM

If washed regularly the bright metalwork on the vehicle should remain untarnished. However if tarnishing does occur it can be removed by using a 'chrome cleaner'. Harsh abrasives, such as wire wool, should never be used for cleaning brightwork as they are likely to score the metal surface.

Car washing !

POLISHING

A good quality wax polish covering the entire painted surface of the car will form an extremely effective shield against weather conditions and also improve the appearance of the vehicle. It is therefore advisable to 'wax' the car occasionally.

General Procedure

(1) Ensure that the bodywork is clean and dry.
(2) In some instances, owing to neglect, it may be necessary to clean the paintwork with a special cleaner (Haze remover) prior to waxing.
(3) It is usual to apply the wax and polish to quite small sections at a time (for example, a door panel).
(4) It is not advisable to wax (or wash) a car in strong sunlight.

HAZE REMOVER

As the name implies, this is a cleaner which will remove the grime and traffic film which may have built up over a period as a result of neglect. It does however remove a very thin surface layer of paint and should not be used too often. Haze remover is sold under a variety of names such as 'cutting', or 'brazing' compound.

UNDERBODY

Any places under the body (and under the wings) where mud is likely to be trapped, should be pressure-hosed from time to time to prevent corrosion developing. The underbody should also be checked for loss of paint or protective coating, and any affected areas touched in.

CAR INTERIOR

The carpets should be cleaned with a stiff brush or vacuum cleaner, preferably before washing the car. If it is possible the carpets and underfelt should be removed occasionally in order to check the floor area for dampness.

Seats and interior door panels etc. can be cleaned by using a proprietary upholstery cleaner, however, unless these areas have been neglected, hot soapy water is usually adequate for this purpose. (Note: Leather upholstery should be treated with 'hide food'.)

CARE OF PAINTWORK

It is an opportune time when cleaning a vehicle to check for chipped and scratched paintwork. Minor blemishes can be rectified by 'touch-up' paint. Before painting it is however necessary to remove any rust and thoroughly clean the affected areas.

76

EXERCISE 13. CLEANING and VALETING

Questions

(1) To cover the work outlined on the previous two pages, the vehicle owner should possess certain equipment and materials. Draw up a list of equipment and materials which would be suitable for this purpose.

(2) Why is it beneficial to wax-polish a car?

(3) Outline the procedure for cleaning a vehicle which has been used for a long period (for example six months) without being washed.

(4) Unless mud is prevented from accumulating in 'traps' under the wings and elsewhere corrosion is likely to occur. What do you think is the main reason for this?

(5) What areas on a car bodywork are most likely to suffer from chipped and scratched paintwork? Give reasons for your answer.

Vehicle Investigation (State make, model and year of vehicle)

(a) Examine a vehicle and complete a table similar to that shown below.

Examined	Faults	Possible causes	Treatment necessary
Exterior paintwork and brightwork			
Inside the car			
Body underside			

(b) Try applying wax polish to three small areas of a dirty car; first without any preparation; secondly after simply washing; and thirdly after using Haze remover. Compare the results.

(c) Using paraffin and a brush (or any proprietary cleaner) but NOT petrol, clean the engine compartment of a car to make it as near new-looking as possible.

Project Work

(i) By experiment, determine which type of car polish you consider gives best results in respect of (i) ease of application; (ii) longevity of results; (iii) economy of cost.

(ii) Some cars make use of stainless steels and irons instead of chromium plate. Give examples of such cars and say which parts are of stainless materials. State the advantages and disadvantages of this practice (for example, in Rolls-Royce radiators).

(iii) The paint used on car bodies is very different from the paint used for such things as houses. Find out about modern car body paints and their special features.

SECTION 14

MOTOR CYCLES AND SCOOTERS

GENERAL LAYOUT (Motor Cycles)

Motor cycles consist of a basic frame, to which are attached all the major components—these are:

Engine, transmission, fuel tank, steering gear, seat, suspension and wheels.

The frame construction is somewhat similar to that of an ordinary pedal bicycle. It consists of steel tubes which are 'brazed' into cast or forged lugs.

A typical motor cycle frame is shown below:

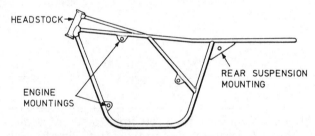

A typical example of modern motor cycle layout is this:

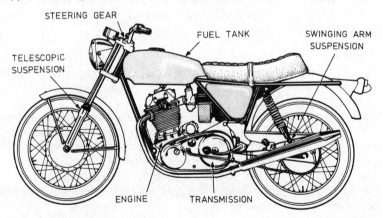

ENGINE

The principle of operation of the motor cycle engine is exactly the same as that for a motor car engine; that is, the internal combustion petrol engine of either the two-stroke or four-stroke design. However many motor cycles employ two-stroke engines, whereas the two-stroke is used comparatively rarely in motor cars.

The main differences between motor cycle engines and car engines are:

(a) For obvious reasons, motor cycle engines are smaller.
(b) The majority of motor cycle engines employ only one or two cylinders.
(c) Almost all motor cycle engines are air-cooled.

Types

Motor cycle engines, like those for cars, are usually classified by their capacity (expressed in cubic centimetres cc) and the number of cylinders. For example:

125 cc single	250 cc single	500 cc single
500 cc twin	650 cc twin	

A typical engine assembly, with the main components exposed is shown below.

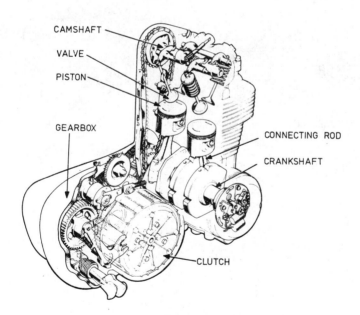

Cylinder Arrangement

On a single cylinder machine the cylinder can either be positioned vertically; inclined towards the front; or even horizontally. When two cylinders are employed, they are usually arranged side by side. Exceptions to this are cylinders which are 'horizontally opposed' across the frame, or cylinders arranged in a 'vee' formation in line with the frame. Some modern motor cycles employ three cylinder engines in which the cylinders are arranged side by side; and in certain instances four cylinder engines are used.

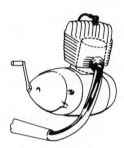

INCLINED SINGLE

VERTICAL TWIN

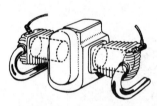

HORIZONTALLY OPPOSED TWIN

THREE CYLINDERS

EXERCISE 14A.   MOTORCYCLES, LAYOUT and ENGINE

Questions

(1) Why do motor cycles commonly have heavily finned sumps?
(2) Make a simple line diagram to show the crankshaft shape on a vertical twin, four-stroke engine.
(3) What are the advantages of air-cooling that makes it so very suitable for motor cycles?
(4) For what reasons are multi-cylinder engines used in preference to one large single-cylinder?
(5) What are the advantages of two-stroke engines for motor cycles?

Vehicle Investigation (State make, model and year of machine)

(a) Examine a machine and list its most important features including engine type, capacity, number of cylinders and so on.
(b) Sketch the layout of the frame and show the places at which it has been brazed. Indicate any parts that require repair.
(c) List all the parts that require regular lubrication and state the appropriate lubricant in each case.
(d) Sketch the engine layout and indicate how it is mounted in the frame.

Project Work

 (i) List as many as you can of foreign-built motor cycles currently for sale in Britain. Explain what you consider are the reasons for their popularity.
(ii) What are the very latest developments in motor cycle design and what do you consider are their future prospects?

TRANSMISSION

The drive from the engine to the rear wheel is transmitted through a clutch and gearbox. Both these components serving the same purpose as they do in a motor car. On most motor cycles chains (rather than shafts) are used to transmit the drive; firstly the *primary chain* from the engine to the clutch; and secondly the *secondary chain* from the gearbox to the rear wheel.

Gear Ratios

The average 'primary' reduction ratio is in the region of 2 :1, whilst the average 'secondary' reduction ratio is about 3 : 1. If therefore the top gear ratio in the gearbox is 1 : 1, the overall transmission ratio when in top gear is 6 : 1 (2 × 3). This means that the engine will rotate six times faster than the rear wheel when the machine is travelling in top gear.

Chain Arrangement (Plan View)

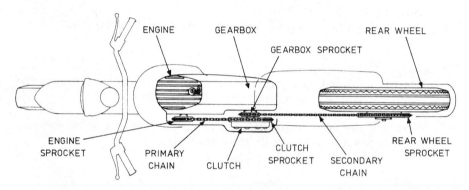

This drawing shows a typical transmission arrangement for a motor cycle, whilst the majority of motor cycles follow this general pattern, some machines transmit the primary drive through gears. As can be seen from the drawing, the input and output are at the same end of the gearbox, somewhat different therefore from the motor car gearbox.

Principle of Operation (Gearbox)

The drawings show how the drive passes into and out of the gearbox, although to aid understanding the drawings have been simplified and the gearbox shown with only two speeds. The engine clutch (shown on left) used in conjunction with the gearbox is normally of the multi-plate friction type, but nevertheless the principle of operation is basically similar to the type described on motor cars. In this instance however, the friction linings connect the clutch sprocket to the gearbox input shaft.

Neutral:

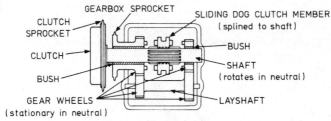

Low gear:

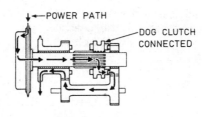

High gear:

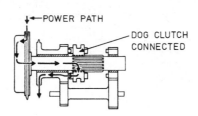

SUSPENSION

The suspension units at both front and rear on most motor cycles are of the telescopic type (sometimes referred to as 'teledraulics'). Each unit can be simply described as being a telescopic damper (described under motor vehicle suspension) which incorporates a coil spring, thus forming an extremely neat and compact suspension assembly.

Front Suspension

The front suspension units are attached to the steering stem and in this way form the front 'forks' which locate the axle and wheel assembly.

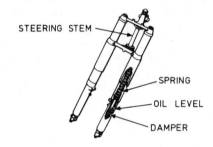

Rear Suspension

Rear suspension is normally of the 'swinging arm' type. In this type the rear 'forks' which carry the wheel and axle assembly are pivoted on a bearing immediately behind the gearbox, with the suspension units positioned as shown in the drawing below.

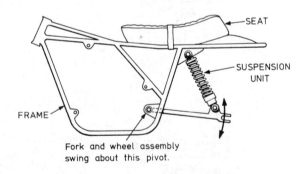

Fork and wheel assembly
swing about this pivot.

EXERCISE 14B.   MOTORCYCLES, TRANSMISSION and SUSPENSION

Questions

(1) Why are the input and output shafts, shown on pages 82 and 83, both on the same side of the gearbox?
(2) Why does a motor cycle require two chains?
(3) What form of final drive to the rear wheel may be used, other than chain drive?
(4) In what way would the use of a smaller than usual rear road wheel affect the gear ratio?
(5) Why is it that motor cycles are not made with front wheel drive?

Vehicle Investigation (State make, model and year of machine)

(a) Find out how the secondary chain tension can be adjusted and show the arrangement by a sketch.
(b) How is the transmission lubricated?
(c) Determine whether or not it is possible to remove the rear wheel without disturbing the drive to it.
(d) With the machine in different gears, determine the numerical ratio of each gear; that is, the number of engine revolutions to one revolution of the driving wheel.

Project Work

(i) Determine the relative merits of chain and shaft drive.
(ii) Motor cycle suspension has improved tremendously since World War II. Trace its development and state the most important changes.

CONTROLS

The rider of a motor cycle must, as well as steering the machine and keeping it upright, be able to operate easily the throttle, clutch, gearchange, brakes and other minor controls. Usually the throttle, clutch and front brake are hand-operated via 'Bowden cables' controlled by levers mounted on the handlebars, whilst the gearchange and rear brake are operated by foot pedals.

The sketches below show the method of operating the clutch and front brake by means of 'Bowden cables' and hand levers:

Clutch Control

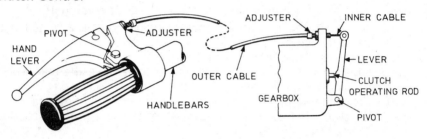

Front Brake Control

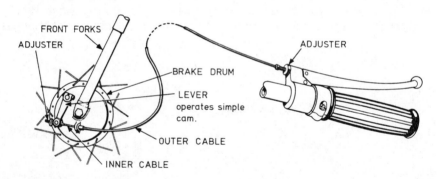

Adjustment

As can be seen from the sketches, the correct amount of 'free play' in the mechanism is maintained by altering the effective length of the outer cable. (This is the same system as commonly used on bicycle brakes.)

Throttle Control

The throttle cable is operated by a device known as a 'twistgrip'. This is usually the right-hand handlegrip which is attached to a tubular sleeve mounted on the handlebar. The throttle cable is attached to the sleeve as shown below. To operate the throttle, the twistgrip is rotated and this action winds the cable around the sleeve, thus opening the throttle.

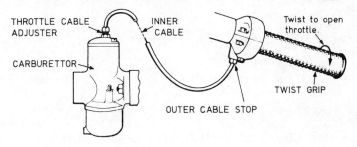

Gearchange

The gearchange foot pedal is clamped to a shaft (protruding from the gearbox) which, when rotated, operates the gear selector mechanism. The movement of the pedal for one type of four-speed gearbox is shown below. After each gearchange has been made the pedal returns to the normal position. The numbers on the drawing only indicate 'direction of pedal movement', not 'pedal positions'.

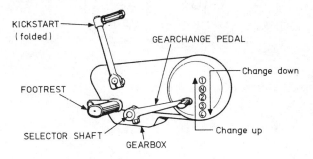

Rear Brake Control

The rear brake is normally rod-operated from a pedal on the opposite side of the machine to the gearchange pedal.

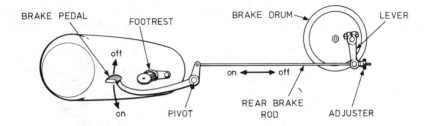

Kick-Start

A 'kick-start' device is normally used to start motor cycle engines. When the kick-start lever is pushed down, the engine is rotated. The drive from the kick-start to the engine is transmitted through the gearbox and clutch.

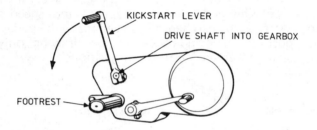

Arrangement of Main Controls

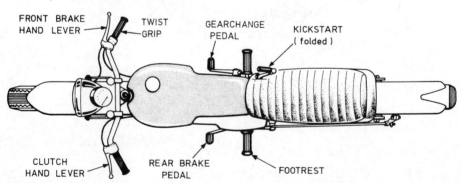

EXERCISE 14C. MOTORCYCLE CONTROLS

Questions

(1) SAFETY: List the desirable clothing and headgear that a motor cycle rider should use.
(2) What is the sequence of rider operations when making a gearchange?
(3) Why are the clutch and front brake almost invariably hand-operated?
(4) Why must the front brake always be cable-operated whilst the rear brake is virtually always rod-operated?
(5) Why are all motor cycles fitted with an on/off petrol tap?

Vehicle Investigation (State make, model and year of machine)

(a) What is the procedure for starting the engine?
(b) Make a sketch to illustrate the direction of gear lever movement through the various gears.
(c) What maintenance do the various hand- and foot-operated controls require?
(d) What adjustments are there to help provide for driver comfort? (Note: Inspect handlebar fixing, footrests and footlevers, etc.)
(e) What maintenance does the suspension require?

Project Work

 (i) In what ways do the controls on motor cycles used for racing and scrambling differ from those used on ordinary machines?
(ii) Study the layout of controls of machines from different countries and determine how their layouts and operation differ.

SCOOTERS

The basic mechanics of the scooter are much the same as those of the motor cycle, the major difference being in the layout of the machine. A scooter is designed to give an alternative driving position to the one adopted on motor cycles, i.e. the rider sits with his legs in front as if sitting on a chair, whereas on a motor cycle the rider straddles the machine in the same way as a rider straddles a horse.

To give this driving position, the fuel tank, engine and transmission, etc. are usually positioned beneath and to the rear of the driver, below the seat. A typical scooter showing the arrangement of the main components is shown below.

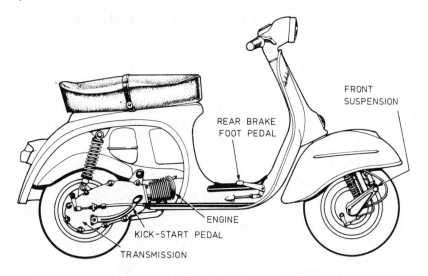

FRONT
SUSPENSION

REAR BRAKE
FOOT PEDAL

ENGINE

KICK-START PEDAL

TRANSMISSION

Owing to the compact arrangement of the engine and transmission some scooters employ a gear drive to the rear wheel rather than a chain as used on most motor cycles.

EXERCISE 14D. SCOOTERS

Questions

(1) What are the advantages of a scooter when compared with a motor cycle?

(2) In what ways do the major controls on a scooter and motor cycle differ?

(3) Why must the mechanical components be largely bunched together below the driver's seat?

(4) Why are the road wheels so much smaller than those of motor cycles?

Vehicle Investigation (State make, model and year of machine)

(a) Examine a scooter and make sketches to show the arrangement of the engine and transmission. Show how the assembly is attached to the frame.

(b) How is the drive transmitted from the engine to the rear wheel?

(c) Sketch the front suspension arrangement and briefly explain how it differs from that employed on most motor cycles.

(d) Describe with the aid of sketches, the method of selecting the gears.

(e) Sketch the basic frame layout and state how it differs substantially from that of a motor cycle.

Project Work

(i) By making a survey of a number of modern motor cycles and scooters, compile a table which gives a comparison of prices, engine types, performance figures (maximum speed, fuel consumption, etc.).

(ii) By studying modern scooters, determine what you consider to be likely future design trends.